The Principle

Albert Einstein and H. Minkowski

Alpha Editions

This edition published in 2024

ISBN 9789362518132

Design and Setting By
Alpha Editions
www.alphaedis.com
Email - info@alphaedis.com

Contents

HISTORICAL INTRODUCTION

Lord Kelvin writing in 1893, in his preface to the English edition of Hertz's Researches on Electric Waves, says "many workers and many thinkers have helped to build up the nineteenth century school of *plenum*, one ether for light, heat, electricity, magnetism; and the German and English volumes containing Hertz's electrical papers, given to the world in the last decade of the century, will be a permanent monument of the splendid consummation now realised."

Ten years later, in 1905, we find Einstein declaring that "the ether will be proved to be superfluous." At first sight the revolution in scientific thought brought about in the course of a single decade appears to be almost too violent. A more careful even though a rapid review of the subject will, however, show how the Theory of Relativity gradually became a historical necessity.

Towards the beginning of the nineteenth century, the luminiferous ether came into prominence as a result of the brilliant successes of the wave theory in the hands of Young and Fresnel. In its stationary aspect the elastic solid ether was the outcome of the search for a medium in which the light waves may "undulate." This stationary ether, as shown by Young, also afforded a satisfactory explanation of astronomical aberration. But its very success gave rise to a host of new questions all bearing on the central problem of relative motion of ether and matter.

Arago's prism experiment.—The refractive index of a glass prism depends on the incident velocity of light outside the prism and its velocity inside the prism after refraction. On Fresnel's fixed ether hypothesis, the incident light waves are situated in the stationary ether outside the prism and move with velocity c with respect to the ether. If the prism moves with a velocity u with respect to this fixed ether, then the incident velocity of light with respect to the prism should be $c + u$. Thus the refractive index of the glass prism should depend on u the absolute velocity of the prism, *i.e.*, its velocity with respect to the fixed ether. Arago performed the experiment in 1819, but failed to detect the expected change.

Airy-Boscovitch water-telescope experiment.—Boscovitch had still earlier in 1766, raised the very important question of the dependence of aberration on the refractive index of the medium filling the telescope. Aberration depends on the difference in the velocity of light outside the telescope and its velocity inside the telescope. If the latter velocity changes owing to a change in the medium filling the telescope, aberration itself should change, that is, aberration should depend on the nature of the medium.

Airy, in 1871 filled up a telescope with water—but failed to detect any change in the aberration. Thus we get both in the case of Arago prism experiment and Airy-Boscovitch water-telescope experiment, the very startling result that optical effects in a moving medium seem to be quite independent of the velocity of the medium with respect to Fresnel's stationary ether.

Fresnel's convection coefficient $k = 1 - 1/\mu^2$.—Possibly some form of compensation is taking place. Working on this hypothesis, Fresnel offered his famous ether convection theory. According to Fresnel, the presence of matter implies a definite condensation of ether within the region occupied by matter. This "condensed" or excess portion of ether is supposed to be carried away with its own piece of moving matter. It should be observed that only the "excess" portion is carried away, while the rest remains as stagnant as ever. A complete convection of the "excess" ether ϱ' with the full velocity u is optically equivalent to a partial convection of the total ether ϱ, with only a fraction of the velocity $k. u$. Fresnel showed that if this convection coefficient k is $1 - 1/\mu^2$ (μ being the refractive index of the prism), then the velocity of light after refraction within the moving prism would be altered to just such extent as would make the refractive index of the moving prism quite independent of its "absolute" velocity u. The non-dependence of aberration on the "absolute" velocity u, is also very easily explained with the help of this Fresnelian convection-coefficient k.

Stokes' viscous ether.—It should be remembered, however, that Fresnel's stationary ether is absolutely fixed and is not at all disturbed by the motion of matter through it. In this respect Fresnelian ether cannot be said to behave in any respectable physical fashion, and this led Stokes, in 1845-46, to construct a more material type of medium. Stokes assumed that viscous motion ensues near the surface of separation of ether and moving matter, while at sufficiently distant regions the ether remains wholly undisturbed. He showed how such a viscous ether would explain aberration if all motion in it were differentially irrotational. But in order to explain the null Arago effect, Stokes was compelled to assume the convection hypothesis of Fresnel with an identical numerical value for k, namely $1 - 1/\mu^2$. Thus the prestige of the Fresnelian convection-coefficient was enhanced, if anything, by the theoretical investigations of Stokes.

Fizeau's experiment.—Soon after, in 1851, it received direct experimental confirmation in a brilliant piece of work by Fizeau.

If a divided beam of light is re-united after passing through two adjacent cylinders filled with water, ordinary interference fringes will be produced. If the water in one of the cylinders is now made to flow, the "condensed" ether within the flowing water would be convected and would produce a shift in the interference fringes. The shift actually observed agreed very well with a value of $k = 1 - 1/\mu^2$. The Fresnelian convection-coefficient now became firmly established as a consequence of a direct positive effect. On the other hand, the negative evidences in favour of the convection-coefficient had also multiplied. Mascart, Hoek, Maxwell and others sought for definite changes in different optical effects induced by the motion of the earth relative to the stationary ether. But all such attempts failed to reveal the slightest trace of any optical disturbance due to the "absolute" velocity of the earth, thus proving conclusively that all the different optical effects shared in the general compensation arising out of the Fresnelian convection of the excess ether. It must be carefully noted that the Fresnelian convection-coefficient implicitly assumes the existence of a fixed ether (Fresnel) or at least a wholly stagnant medium at sufficiently distant regions (Stokes), with reference to which alone a convection velocity can have any significance. Thus the convection-coefficient implying some type of a stationary or viscous, yet nevertheless "absolute" ether, succeeded in explaining satisfactorily all known optical facts down to 1880.

Michelson-Morley Experiment.—In 1881, Michelson and Morley performed their classical experiments which undermined the whole structure of the old ether theory and thus served to introduce the new theory of relativity. The fundamental idea underlying this experiment is quite simple. In all old experiments the velocity of light situated in free ether was compared with the velocity of waves actually situated in a piece of moving matter and presumably carried away by it. The compensatory effect of the Fresnelian convection of ether afforded a satisfactory explanation of all negative results.

In the Michelson-Morley experiment the arrangement is quite different. If there is a definite gap in a rigid body, light waves situated in free ether will take a definite time in crossing the gap. If the rigid platform carrying the gap is set in motion with respect to the ether in the direction of light propagation, light waves (which are even now situated in free ether) should presumably take a longer time to cross the gap.

We cannot do better than quote Eddington's description of this famous experiment. "The principle of the experiment may be illustrated by considering a swimmer in a river. It is easily realized that it takes longer to swim to a point 50 yards up-stream and back than to a point 50 yards across-stream and back. If the earth is moving through the ether there is a river of ether flowing through the laboratory, and a wave of light may be compared to a swimmer travelling with constant velocity relative to the current. If, then, we divide a beam of light into two parts, and send one-half swimming up the stream for a certain distance and then (by a mirror) back to the starting point, and send the other half an equal distance across stream and back, the across-stream beam should arrive back first.

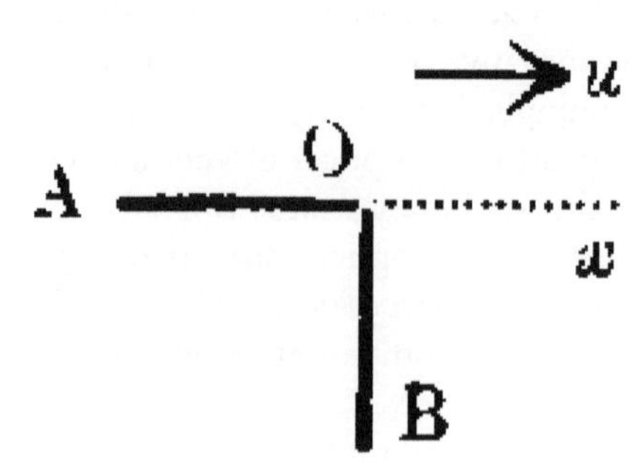

Let the ether be flowing relative to the apparatus with velocity *u* in the direction O*x*, and let OA, OB, be the two arms of the apparatus of equal length *l*, OA being placed up-stream. Let *c* be the velocity of light. The time for the double journey along OA and back is where a factor greater than unity.

$$t_1 = \frac{l}{c-u} + \frac{l}{c+u} = \frac{2lc}{c^2-u^2} = \frac{2l}{c}\beta^2$$

$$\beta = (1 - u^2/c^2)^{-\frac{1}{2}}$$

For the transverse journey the light must have a component velocity *u* up-stream (relative to the ether) in order to avoid being carried below OB: and since its total velocity is *c*, its component across-stream must be $\sqrt{(c^2 - u^2)}$, the time for the double journey OB is accordingly so that $t_1 > t_2$.

$$t_2 = \frac{2a}{\sqrt{(c^2 - u^2)}} = \frac{2a}{c}\beta;$$

But when the experiment was tried, it was found that both parts of the beam took the same time, as tested by the interference bands produced."

After a most careful series of observations, Michelson and Morley failed to detect the slightest trace of any effect due to earth's motion through ether.

The Michelson-Morley experiment seems to show that there is no relative motion of ether and matter. Fresnel's stagnant ether requires a relative velocity of—u. Thus Michelson and Morley themselves thought at first that their experiment confirmed Stokes' viscous ether, in which no relative motion can ensue on account of the absence of slipping of ether at the surface of separation. But even on Stokes' theory this viscous flow of ether would fall off at a very rapid rate as we recede from the surface of separation. Michelson and Morley repeated their experiment at different heights from the surface of the earth, but invariably obtained the same negative results, thus failing to confirm Stokes' theory of viscous flow.

Lodge's experiment.—Further, in 1893, Lodge performed his rotating sphere experiment which showed complete absence of any viscous flow of ether due to moving masses of matter. A divided beam of light, after repeated reflections within a very narrow gap between two massive hemispheres, was allowed to re-unite and thus produce interference bands. When the two hemispheres are set rotating, it is conceivable that the ether in the gap would be disturbed due to viscous flow, and any such flow would be immediately detected by a disturbance of the interference bands. But actual observation failed to detect the slightest disturbance of the ether in the gap, due to the motion of the hemispheres. Lodge's experiment thus seems to show a complete absence of any viscous flow of ether.

Apart from these experimental discrepancies, grave theoretical objections were urged against a viscous ether. Stokes himself had shown that his ether must be incompressible and all motion in it differentially irrotational, at the same time there should be absolutely no slipping at the surface of separation. Now all these conditions cannot be simultaneously satisfied for any conceivable material medium without certain very special and arbitrary assumptions. Thus Stokes' ether failed to satisfy the very motive which had led Stokes to formulate it, namely, the desirability of constructing a "physical" medium. Planck offered modified forms of Stokes' theory which seemed capable of being reconciled with the Michelson-Morley experiment, but required very special assumptions. The very complexity and the very arbitrariness of these assumptions prevented Planck's ether from attaining any degree of practical importance in the further development of the subject.

The sole criterion of the value of any scientific theory must ultimately be its capacity for offering a simple, unified, coherent and fruitful description of observed facts. In proportion as a theory becomes complex it loses in usefulness—a theory which is obliged to requisition a whole array of arbitrary assumptions in order to explain special facts is practically worse than useless, as it serves to disjoin, rather than to unite, the several groups of facts. The optical experiments of the last quarter of the nineteenth century showed the impossibility of constructing a simple ether theory, which would be amenable to analytic treatment and would at the same time stimulate further progress. It should be observed that it could scarcely be shown that no logically consistent ether theory was possible; indeed in 1910, H. A. Wilson offered a consistent ether theory which was at least quite neutral with respect to all available optical data. But Wilson's ether is almost wholly negative—its only virtue being that it does not directly contradict observed facts. Neither any direct confirmation nor a direct refutation is possible and it does not throw any light on the various optical phenomena. A theory like this being practically useless stands self-condemned.

We must now consider the problem of relative motion of ether and matter from the point of view of electrical theory. From 1860 the identity of light as an electromagnetic vector became gradually established as a result of the brilliant "displacement current" hypothesis of Clerk Maxwell and his further analytical investigations. The elastic solid ether became gradually transformed into the electromagnetic one. Maxwell succeeded in giving a fairly satisfactory account of all ordinary optical phenomena and little room was left for any serious doubts as regards the general validity of Maxwell's theory. Hertz's researches on electric waves, first carried out in 1886, succeeded in furnishing a strong experimental confirmation of Maxwell's theory. Electric waves behaved generally like light waves of very large wave length.

The orthodox Maxwellian view located the dielectric polarisation in the electromagnetic ether which was merely a transformation of Fresnel's stagnant ether. The magnetic polarisation was looked upon as wholly secondary in origin, being due to the relative motion of the dielectric tubes of polarisation. On this view the Fresnelian convection coefficient comes out to be ½, as shown by J. J. Thomson in 1880, instead of $1 - (1/\mu^2)$ as required by optical experiments. This obviously implies a complete failure to account for all those optical experiments which depend for their satisfactory explanation on the assumption of a value for the convection coefficient equal to $1 - (1/\mu^2)$.

The modifications proposed independently by Hertz and Heaviside fare no better.[1] They postulated the actual medium to be the seat of all electric polarisation and further emphasised the reciprocal relation subsisting between electricity and magnetism, thus making the field equations more symmetrical. On this view the whole of the polarised ether is carried away by the moving medium, and consequently, the convection coefficient naturally becomes unity in this theory, a value quite as discrepant as that obtained on the original Maxwellian assumption.

Thus neither Maxwell's original theory nor its subsequent modifications as developed by Hertz and Heaviside succeeded in obtaining a value for Fresnelian coefficient equal to $1 - (1/\mu^2)$, and consequently stood totally condemned from the optical point of view.

Certain direct electromagnetic experiments involving relative motion of polarised dielectrics were no less conclusive against the generalised theory of Hertz and Heaviside. According to Hertz a moving dielectric would carry away the whole of its electric displacement with it. Hence the electromagnetic effect near the moving dielectric would be proportional to the total electric displacement, that is to K, the specific inductive capacity of the dielectric. In 1901, Blondlot working with a stream of moving gas could not detect any such effect. H. A. Wilson repeated the experiment in an improved form in 1903 and working with ebonite found that the observed effect was proportional to K - 1 instead of to K. For gases K is nearly equal to 1 and hence practically no effect will be observed in their case. This gives a satisfactory explanation of Blondlot's negative results.

Rowland had shown in 1876 that the magnetic force due to a rotating condenser (the dielectric remaining stationary) was proportional to K, the sp. ind. cap. On the other hand, Röntgen found in 1888 the magnetic effect due to a rotating dielectric (the condenser remaining stationary) to be proportional to K - 1, and not to K. Finally Eichenwald in 1903 found that when both condenser and dielectric are rotated together, the effect observed was quite independent of K, a result quite consistent with the two previous experiments. The Rowland effect proportional to K, together with the opposite Röntgen effect proportional to 1 - K, makes the Eichenwald effect independent of K.

All these experiments together with those of Blondlot and Wilson made it clear that the electromagnetic effect due to a moving dielectric was proportional to K - 1, and not to K as required by Hertz's theory. Thus the above group of experiments with moving dielectrics directly contradicted the Hertz-Heaviside theory. The internal discrepancies inherent in the classic ether theory had now become too prominent. It was clear that the ether concept had finally outgrown its usefulness. The observed facts had become too contradictory and too heterogeneous to be reduced to an organised whole with the help of the ether concept alone. Radical departures from the classical theory had become absolutely necessary.

There were several outstanding difficulties in connection with anomalous dispersion, selective reflection and selective absorption which could not be satisfactory explained in the classic electromagnetic theory. It was evident that the assumption of some kind of discreteness in the optical medium had become inevitable. Such an assumption naturally gave rise to an atomic theory of electricity, namely, the modern electron theory. Lorentz had postulated the existence of electrons so early as 1878, but it was not until some years later that the electron theory became firmly established on a satisfactory basis.

Lorentz assumed that a moving dielectric merely carried away its own "polarisation doublets," which on his theory gave rise to the induced field proportional to $K - 1$. The field near a moving dielectric is naturally proportional to $K - 1$ and not to K. Lorentz's theory thus gave a satisfactory explanation of all those experiments with moving dielectrics which required effects proportional to $K - 1$. Lorentz further succeeded in obtaining a value for the Fresnelian convection coefficient equal to $1 - 1/\mu^2$, the exact value required by all optical experiments of the moving type.

We must now go back to Michelson and Morley's experiment. We have seen that both parts of the beam are situated in free ether; no material medium is involved in any portion of the paths actually traversed by the beam. Consequently no compensation due to Fresnelian convection of ether by moving medium is possible. Thus Fresnelian convection compensation can have no possible application in this case. Yet some marvellous compensation has evidently taken place which has completely masked the "absolute" velocity of the earth.

In Michelson and Morley's experiment, the distance travelled by the beam along OA (that is, in a direction parallel to the motion of the platform) is $2l\beta^2$, while the distance travelled by the beam along OB, perpendicular to the direction of motion of the platform, is $2l\beta$. Yet the most careful experiments showed, as Eddington says, "that both parts of the beam took the same time as tested by the interference bands produced. It would seem that OA and OB could not really have been of the same length; and if OB was of length l, OA must have been of length l/β. The apparatus was now rotated through 90°, so that OB became the up-stream. The time for the two journeys was again the same, so that 0B must now be the shorter length. The plain meaning of the experiment is that both arms have a length l when placed along Oy (perpendicular to the direction of motion), and automatically contract to a length l/β, when placed along Ox (parallel to the direction of motion). This explanation was first given by Fitz-Gerald."

This Fitz-Gerald contraction, startling enough in itself, does not suffice. Assuming this contraction to be a real one, the distance travelled with respect to the ether is $2l\beta$ and the time taken for this journey is $2l\beta/c$. But the distance travelled with respect to the platform is always $2l$. Hence the velocity of light with respect to the platform is

$$2l / \frac{2l\beta}{c} = c/\beta$$

a variable quantity depending on the "absolute" velocity of the platform. But no trace of such an effect has ever been found. The velocity of light is always found to be quite independent of the velocity of the platform. The present difficulty cannot be solved by any further alteration in the measure of space. The only recourse left open is to alter the measure of time as well, that is, to adopt the concept of "local time." If a moving clock goes slower so that one 'real' second becomes $1/\beta$ second as measured in the moving system, the velocity of light relative to the platform will always remain c. We must adopt two very startling hypotheses, namely, the Fitz-Gerald contraction and the concept of "local time," in order to give a satisfactory explanation of the Michelson-Morley experiment.

These results were already reached by Lorentz in the course of further developments of his electron theory. Lorentz used a special set of transformation equations[2] for time which implicitly introduced the concept of local time. But he himself failed to attach any special significance to it, and looked upon it rather as a mere mathematical artifice like imaginary quantities in analysis or the circle at infinity in projective geometry. The originality of Einstein at this stage consists in his successful physical interpretation of these results, and viewing them as the coherent organised consequences of a single general principle. Lorentz established the Relativity Theorem[3] (consisting merely of a set of transformation equations) while Einstein generalised it into a Universal Principle. In addition Einstein introduced fundamentally new concepts of space and time, which served to destroy old fetishes and demanded a wholesale revision of scientific concepts and thus opened up new possibilities in the synthetic unification of natural processes.

Newton had framed his laws of motion in such a way as to make them quite independent of the absolute velocity of the earth. Uniform relative motion of ether and matter could not be detected with the help of dynamical laws. According to Einstein neither could it be detected with the help of optical or electromagnetic experiments. Thus the Einsteinian Principle of Relativity asserts that all physical laws are independent of the 'absolute' velocity of an observer.

For different systems, the *form* of all physical laws is conserved. If we chose the velocity of light[4] to be the fundamental unit of measurement for all observers (that is, assume the constancy of the velocity of light in all systems) we can establish a *metric* "one-one" correspondence between any two observed systems, such correspondence depending only the *relative* velocity of the two systems. Einstein's Relativity is thus merely the consistent logical application of the well known physical principle that we can know nothing but *relative* motion. In this sense it is a further extension of Newtonian Relativity.

On this interpretation, the Lorentz-Fitzgerald contraction and "local time" lose their arbitrary character. Space and time as measured by two different observers are naturally diverse, and the difference depends only on their relative motion. Both are equally valid; they are merely different descriptions of the same physical reality. This is essentially the point of view adopted by Minkowski. He considers time itself to be one of the co-ordinate axes, and in his four-dimensional world, that is in the space-time reality, relative motion is reduced to a rotation of the axes of reference. Thus, the diversity in the measurement of lengths and temporal rates is merely due to the static difference in the "frame-work" of the different observers.

The above theory of Relativity absorbed practically the whole of the electromagnetic theory based on the Maxwell-Lorentz system of field equations. It combined all the advantages of classic Maxwellian theory together with an electronic hypothesis. The Lorentz assumption of polarisation doublets had furnished a satisfactory explanation of the Fresnelian convection of ether, but in the new theory this is deduced merely as a consequence of the altered concept of relative velocity. In addition, the theory of Relativity accepted the results of Michelson and Morley's experiments as a definite principle, namely, the principle of the constancy of the velocity of light, so that there was nothing left for explanation in the Michelson-Morley experiment. But even more than all this, it established a single general principle which served to connect together in a simple coherent and fruitful manner the known facts of Physics.

The theory of Relativity received direct experimental confirmation in several directions. Repeated attempts were made to detect the Lorentz-Fitzgerald contraction. Any ordinary physical contraction will usually have observable physical results; for example, the total electrical resistance of a conductor will diminish. Trouton and Noble, Trouton and Rankine, Rayleigh and Brace, and others employed a variety of different methods to detect the Lorentz-Fitzgerald contraction, but invariably with the same negative results. *Whether there is an ether or not, uniform velocity with respect to it can never be detected.* This does not prove that there is no such thing as an ether but certainly does render the ether entirely superfluous. Universal compensation is due to a change in local units of length and time, or rather, being merely different descriptions of the same reality, there is no compensation at all.

There was another group of observed phenomena which could scarcely be fitted into a Newtonian scheme of dynamics without doing violence to it. The experimental work of Kaufmann, in 1901, made it abundantly clear that the "mass" of an electron depended on its velocity. So early as 1881, J. J. Thomson had shown that the inertia of a charged particle increased with its velocity. Abraham now deduced a formula for the variation of mass with velocity, on the hypothesis that an electron always remained a *rigid* sphere. Lorentz proceeded on the assumption that the electron shared in the Lorentz-Fitzgerald contraction and obtained a totally different formula. A very careful series of measurements carried out independently by Bücherer, Wolz, Hupka and finally Neumann in 1913, decided conclusively in favour of the Lorentz formula. This "contractile" formula follows immediately as a direct consequence of the new Theory of Relativity, without any assumption as regards the electrical origin of inertia. Thus the complete agreement of experimental facts with the predictions of the new theory must be considered as confirming it as a principle which goes even beyond the electron itself. The greatest triumph of this new theory consists, indeed, in the fact that a large number of results, which had formerly required all kinds of special hypotheses for their explanation, are now deduced very simply as inevitable consequences of one single general principle.

We have now traced the history of the development of the restricted or special theory of Relativity, which is mainly concerned with optical and electrical phenomena. It was first offered by Einstein in 1905. Ten years later, Einstein formulated his second theory, the Generalised Principle of Relativity. This new theory is mainly a theory of gravitation and has very little connection with optics and electricity. In one sense, the second theory is indeed a further generalisation of the restricted principle, but the former does not really contain the latter as a special case.

Einstein's first theory is restricted in the sense that it only refers to uniform rectilinear motion and has no application to any kind of accelerated movements. Einstein in his second theory extends the Relativity Principle to cases of accelerated motion. If Relativity is to be universally true, then even accelerated motion must be merely *relative motion between matter and matter*. Hence the Generalised Principle of Relativity asserts that "absolute" motion cannot be detected even with the help of gravitational laws.

All movements must be referred to definite sets of co-ordinate axes. If there is any change of axes, the numerical magnitude of the movements will also change. But according to Newtonian dynamics, such alteration in physical movements can only be due to the effect of certain forces in the field.[5] Thus any change of axes will introduce new "geometrical" forces in the field which are quite independent of the nature of the body acted on. Gravitational forces also have this same remarkable property, and gravitation itself may be of essentially the same nature as these "geometrical" forces introduced by a change of axes. This leads to Einstein's famous Principle of Equivalence. *A gravitational field of force is strictly equivalent to one introduced by a transformation of co-ordinates and no possible experiment can distinguish between the two.*

Thus it may become possible to "transform away" gravitational effects, at least for sufficiently small regions of space, by referring all movements to a new set of axes. This new "framework" may of course have all kinds of very complicated movements when referred to the old Galilean or "rectangular unaccelerated system of co-ordinates."

But there is no reason why we should look upon the Galilean system as more fundamental than any other. If it is found simpler to refer all motion in a gravitational field to a special set of co-ordinates, we may certainly look upon this special "framework" (at least for the particular region concerned), to be more fundamental and more natural. We may, still more simply, identify this particular framework with the special local properties of space in that region. That is, we can look upon the effects of a gravitational field as simply due to the local properties of space and time itself. The very presence of matter implies a modification of the characteristics of space and time in its neighbourhood. As Eddington says "matter does not cause the curvature of space-time. It is the curvature. Just as light does not cause electromagnetic oscillations; it is the oscillations."

We may look upon this from a slightly different point of view. The General Principle of Relativity asserts that all motion is merely relative motion between matter and matter, and as all movements must be referred to definite sets of co-ordinates, the ground of any possible framework must ultimately be material in character. It *is* convenient to take the matter actually present in a field as the fundamental ground of our framework. If this is done, the special characteristics of our framework would naturally depend on the actual distribution of matter in the field. But physical space and time is completely defined by the "framework." In other words the "framework" itself *is* space and time. Hence we see how *physical* space and time is actually defined by the local distribution of matter.

There are certain magnitudes which remain constant by any change of axes. In ordinary geometry distance between two points is one such magnitude; so that $\delta x^2 + \delta y^2 + \delta z^2$ is an invariant. In the restricted theory of light, the principle of constancy of light velocity demands that $\delta x^2 + \delta y^2 + \delta z^2 - c^2\delta t^2$ should remain constant.

The *separation ds* of adjacent events is defined by $ds^2 = -dx^2 - dy^2 - dz^2 + c^2dt^2$. It is an extension of the notion of distance and this is the new invariant. Now if x, y, z, t are transformed to any set of new variables x_1, x_2, x_3, x_4, we shall get a quadratic expression for where the *g*'s are functions of x_1, x_2, x_3, x_4 depending on the transformation.

$$ds^2 = g_{11}x_1^2 + 2g_{12}x_1x_2 + \dots = \sum g_{ij}x_ix_j$$

The special properties of space and time in any region are defined by these *g*'s which are themselves determined by the actual distribution of matter in the locality. Thus from the Newtonian point of view, these *g*'s represent the gravitational effect of matter while from the Relativity stand-point, these merely define the non-Newtonian (and incidentally non-Euclidean) space in the neighbourhood of matter.

We have seen that Einstein's theory requires local curvature of space-time in the neighbourhood of matter. Such altered characteristics of space and time give a satisfactory explanation of an outstanding discrepancy in the observed advance of perihelion of Mercury. The large discordance is almost completely removed by Einstein's theory.

Again, in an intense gravitational field, a beam of light will be affected by the local curvature of space, so that to an observer who is referring all phenomena to a Newtonian system, the beam of light will appear to deviate from its path along an Euclidean straight line.

This famous prediction of Einstein about the deflection of a beam of light by the sun's gravitational field was tested during the total solar eclipse of

May, 1919. The observed deflection is decisively in favour of the Generalised Theory of Relativity.

It should be noted however that the velocity of light itself would decrease in a gravitational field. This may appear at first sight to be a violation of the principle of constancy of light-velocity. But when we remember that the Special Theory is explicitly *restricted* to the case of unaccelerated motion, the difficulty vanishes. In the absence of a gravitational field, that is in any unaccelerated system, the velocity of light will always remain constant. Thus the validity of the Special Theory is completely preserved within its own *restricted* field.

Einstein has proposed a third crucial test. He has predicted a shift of spectral lines towards the red, due to an intense gravitational potential. Experimental difficulties are very considerable here, as the shift of spectral lines is a complex phenomenon. Evidence is conflicting and nothing conclusive can yet be asserted. Einstein thought that a gravitational displacement of the Fraunhofer lines is a necessary and fundamental condition for the acceptance of his theory. But Eddington has pointed out that even if this test fails, the logical conclusion would seem to be that while Einstein's law of gravitation is true for matter in bulk, it is not true for such small material systems as atomic oscillator.

Conclusion

From the conceptual stand-point there are several important consequences of the Generalised or Gravitational Theory of Relativity. Physical space-time is perceived to be intimately connected with the actual local distribution of matter. Euclid-Newtonian space-time is *not* the actual space-time of Physics, simply because the former completely neglects the actual presence of matter. Euclid-Newtonian continuum is merely an abstraction, while physical space-*time is the actual framework which has some definite curvature due to the presence of matter. Gravitational Theory of Relativity thus brings out clearly the fundamental distinction between actual physical space-time (which is non-isotropic and non-Euclid-Newtonian) on one hand and the abstract Euclid-Newtonian continuum (which is homogeneous, isotropic and a purely intellectual construction) on the other.

The measurements of the rotation of the earth reveals a fundamental framework which may be called the "inertial framework." This constitutes the actual physical universe. This universe approaches Galilean space-time at a great distance from matter.

The properties of this physical universe may be referred to some world-distribution of matter or the "inertial framework" may be constructed by a suitable modification of the law of gravitation itself. In Einstein's theory the actual curvature of the "inertial framework" is referred to vast quantities of undetected world-matter. It has interesting consequences. The dimensions of Einsteinian universe would depend on the quantity of matter in it; it would vanish to a point in the total absence of matter. Then again curvature depends on the quantity of matter, and hence in the presence of a sufficient quantity of matter space-time may curve round and close up. Einsteinian universe will then reduce to a finite system without boundaries, like the surface of a sphere. In this "closed up" system, light rays will come to a focus after travelling round the universe and we should see an "anti-sun" (corresponding to the back surface of the sun) at a point in the sky opposite to the real sun. This anti-sun would of course be equally large and equally bright if there is no absorption of light in free space.

In de Sitter's theory, the existence of vast quantities of world-matter is not required. But beyond a definite distance from an observer, time itself stands still, so that to the observer nothing can ever "happen" there. All these theories are still highly speculative in character, but they have certainly extended the scope of theoretical physics to the central problem of the ultimate nature of the universe itself.

One outstanding peculiarity still attaches to the concept of electric force—it is not amenable to any process of being "transformed away" by a suitable change of framework. H. Weyl, it seems, has developed a geometrical theory (in hyper-space) in which no fundamental distinction is made between gravitational and electrical forces.

Einstein's theory connects up the law of gravitation with the laws of motion, and serves to establish a very intimate relationship between matter and physical space-*time. Space, time and matter (or energy) were considered to be the three ultimate elements in Physics. The restricted theory fused space-time into one indissoluble whole. The generalised theory has further synthesised space-time and matter into one fundamental physical reality. Space, time and matter taken separately are more abstractions. Physical reality consists of a synthesis of all three.

P. C. MAHALANOBIS.

Note A.

For example consider a massive particle resting on a circular disc. If we set the disc rotating, a centrifugal force appears in the field. On the other hand, if we transform to a set of rotating axes, we must introduce a centrifugal force in order to correct for the change of axes. This newly introduced centrifugal force is usually looked upon as a mathematical fiction—as "geometrical" rather than physical. The presence of such a geometrical force is usually interpreted as being due to the adoption of a fictitious framework. On the other hand a gravitational force is considered quite real. Thus a fundamental distinction is made between geometrical and gravitational forces.

In the General Theory of Relativity, this fundamental distinction is done away with. The very possibility of distinguishing between geometrical and gravitational forces is denied. All axes of reference may now be regarded as equally valid.

In the Restricted Theory, all "unaccelerated" axes of reference were recognised as equally valid, so that physical laws were made independent of uniform absolute velocity. In the General Theory, physical laws are made independent of "absolute" motion of any kind.

INTRODUCTION.

It is well known that if we attempt to apply Maxwell's electrodynamics, as conceived at the present time, to moving bodies, we are led to asymmetry which does not agree with observed phenomena. Let us think of the mutual action between a magnet and a conductor. The observed phenomena in this case depend only on the relative motion of the conductor and the magnet, while according to the usual conception, a distinction must be made between the cases where the one or the other of the bodies is in motion. If, for example, the magnet moves and the conductor is at rest, then an electric field of certain energy-value is produced in the neighbourhood of the magnet, which excites a current in those parts of the field where a conductor exists. But if the magnet be at rest and the conductor be set in motion, no electric field is produced in the neighbourhood of the magnet, but an electromotive force which corresponds to no energy in itself is produced in the conductor; this causes an electric current of the same magnitude and the same career as the electric force, it being of course assumed that the relative motion in both of these cases is the same.

2. Examples of a similar kind such as the unsuccessful attempt to substantiate the motion of the earth relative to the "Light-medium" lead us to the supposition that not only in mechanics, but also in electrodynamics, no properties of observed facts correspond to a concept of absolute rest; but that for all coordinate systems for which the mechanical equations hold, the equivalent electrodynamical and optical equations hold also, as has already been shown for magnitudes of the first order. In the following we make these assumptions (which we shall subsequently call the Principle of Relativity) and introduce the further assumption,—an assumption which is at the first sight quite irreconcilable with the former one—that light is propagated in vacant space, with a velocity c which is independent of the nature of motion of the emitting body. These two assumptions are quite sufficient to give us a simple and consistent theory of electrodynamics of moving bodies on the basis of the Maxwellian theory for bodies at rest. The introduction of a "Lightäther" will be proved to be superfluous, for according to the conceptions which will be developed, we shall introduce neither a space absolutely at rest, and endowed with special properties, nor shall we associate a velocity-vector with a point in which electro-magnetic processes take place.

3. Like every other theory in electrodynamics, the theory is based on the kinematics of rigid bodies; in the enunciation of every theory, we have to do with relations between rigid bodies (co-ordinate system), clocks, and

electromagnetic processes. An insufficient consideration of these circumstances is the cause of difficulties with which the electrodynamics of moving bodies have to fight at present.

I.—KINEMATICAL PORTION.

§ 1. Definition of Synchronism.

Let us have a co-ordinate system, in which the Newtonian equations hold. For distinguishing this system from another which will be introduced hereafter, we shall always call it "the stationary system."

If a material point be at rest in this system, then its position in this system can be found out by a measuring rod, and can be expressed by the methods of Euclidean Geometry, or in Cartesian co-ordinates.

If we wish to describe the motion of a material point, the values of its coordinates must be expressed as functions of time. It is always to be borne in mind that *such a mathematical definition has a physical sense, only when we have a clear notion of what is meant by time. We have to take into consideration the fact that those of our conceptions, in which time plays a part, are always conceptions of synchronism.* For example, we say that a train arrives here at 7 o'clock; this means that the exact pointing of the little hand of my watch to 7, and the arrival of the train are synchronous events.

It may appear that all difficulties connected with the definition of time can be removed when in place of time, we substitute the position of the little hand of my watch. Such a definition is in fact sufficient, when it is required to define time exclusively for the place at which the clock is stationed. But the definition is not sufficient when it is required to connect by time events taking place at different stations,—or what amounts to the same thing,—to estimate by means of time (zeitlich werten) the occurrence of events, which take place at stations distant from the clock.

Now with regard to this attempt;—the time-estimation of events, we can satisfy ourselves in the following manner. Suppose an observer—who is stationed at the origin of coordinates with the clock—associates a ray of light which comes to him through space, and gives testimony to the event of which the time is to be estimated,—with the corresponding position of the hands of the clock. But such an association has this defect,—it depends on the position of the observer provided with the clock, as we know by experience. We can attain to a more practicable result by the following treatment.

If an observer be stationed at A with a clock, he can estimate the time of events occurring in the immediate neighbourhood of A, by looking for the position of the hands of the clock, which are synchronous with the event. If an observer be stationed at B with a clock,—we should add that the clock is of the same nature as the one at A,—he can estimate the time of events occurring about B. But without further premises, it is not possible to compare, as far as time is concerned, the events at B with the events at A.

We have hitherto an A-time, and a B-time, but no time common to A and B. This last time (*i.e.*, common time) can be defined, if we establish by definition that the time which light requires in travelling from A to B is equivalent to the time which light requires in travelling from B to A. For example, a ray of light proceeds from A at A-time t_A towards B, arrives and is reflected from B at B-time t_B, and returns to A at A-time t'_A. According to the definition, both clocks are synchronous, if

$$t_B - t_A = t'_A - t_B.$$

We assume that this definition of synchronism is possible without involving any inconsistency, for any number of points, therefore the following relations hold:—

1. If the clock at B be synchronous with the clock at A, then the clock at A is synchronous with the clock at B.

2. If the clock at A as well as the clock at B are both synchronous with the clock at C, then the clocks at A and B are synchronous.

Thus with the help of certain physical experiences, we have established what we understand when we speak of clocks at rest at different stations, and synchronous with one another; and thereby we have arrived at a definition of synchronism and time.

In accordance with experience we shall assume that the magnitude

$$\frac{2\,\overline{AB}}{t'_A - t_A} = c$$

where *c* is a universal constant.

We have defined time essentially with a clock at rest in a stationary system. On account of its adaptability to the stationary system, we call the time defined in this way as "time of the stationary system."

§ 2. On the Relativity of Length and Time.

The following reflections are based on the Principle of Relativity and on the Principle of Constancy of the velocity of light, both of which we define in the following way:—

1. The laws according to which the nature of physical systems alter are independent of the manner in which these changes are referred to two co-ordinate systems which have a uniform translators motion relative to each other.

2. Every ray of light moves in the "stationary co-ordinate system" with the same velocity c, the velocity being independent of the condition whether this ray of light is emitted by a body at rest or in motion.[6] Therefore

velocity = Path of Light/Interval of time,

where, by 'interval of time' we mean time as defined in §1.

Let us have a rigid rod at rest; this has a length l, when measured by a measuring rod at rest; we suppose that the axis of the rod is laid along the X-axis of the system at rest, and then a uniform velocity v, parallel to the axis of X, is imparted to it. Let us now enquire about the length of the moving rod; this can be obtained by either of these operations.—

(*a*) The observer provided with the measuring rod moves along with the rod to be measured, and measures by direct superposition the length of the rod:—just as if the observer, the measuring rod, and the rod to be measured were at rest.

(*b*) The observer finds out, by means of clocks placed in a system at rest (the clocks being synchronous as defined in §1), the points of this system where the ends of the rod to be measured occur at a particular time t. The distance between these two points, measured by the previously used measuring rod, this time it being at rest, is a length, which we may call the "length of the rod."

According to the Principle of Relativity, the length found out by the operation *a*), which we may call "the length of the rod in the moving system" is equal to the length l of the rod in the stationary system.

The length which is found out by the second method, may be called '*the length of the moving rod measured from the stationary system.*' This length is to be estimated on the basis of our principle, and *we shall find it to be different from* l.

In the generally recognised kinematics, we silently assume that the lengths defined by these two operations are equal, or in other words, that at an epoch of time t, a moving rigid body is geometrically replaceable by the same body, which can replace it in the condition of rest.

Relativity of Time.

Let us suppose that the two clocks synchronous with the clocks in the system at rest are brought to the ends A, and B of a rod, *i.e.*, the time of the clocks correspond to the time of the stationary system at the points where they happen to arrive; these clocks are therefore synchronous in the stationary system.

We further imagine that there are two observers at the two watches, and moving with them, and that these observers apply the criterion for synchronism to the two clocks. At the time t_A, a ray of light goes out from A, is reflected from B at the time t_B, and arrives back at A at time t'_A. Taking into consideration the principle of, constancy of the velocity of light, we have

$$t_B - t_A = r_{AB}/(c - v),$$

$$\text{and } t'_A - t_B = r_{AB}/(c + v),$$

where r_{AB} is the length of the moving rod, measured in the stationary system. Therefore the observers stationed with the watches will not find the clocks synchronous, though the observer in the stationary system must declare the clocks to be synchronous. We therefore see that we can attach no absolute significance to the concept of synchronism; but two events which are synchronous when viewed from one system, will not be synchronous when viewed from a system moving relatively to this system.

§ 3. Theory of Co-ordinate and Time-Transformation from a stationary system to a system which moves relatively to this with uniform velocity.

Let there be given, in the stationary system two co-ordinate systems, *i.e.*, two series of three mutually perpendicular lines issuing from a point. Let the X-axes of each coincide with one another, and the Y and Z-axes be parallel. Let a rigid measuring rod, and a number of clocks be given to each of the systems, and let the rods and clocks in each be exactly alike each other.

Let the initial point of one of the systems (k) have a constant velocity in the direction of the X-axis of the other which is stationary system K, the motion being also communicated to the rods and clocks in the system (k). Any time t of the stationary system K corresponds to a definite position of the axes of the moving system, which are always parallel to the axes of the stationary system. By t, we always mean the time in the stationary system.

We suppose that the space is measured by the stationary measuring rod placed in the stationary system, as well as by the moving measuring rod placed in the moving system, and we thus obtain the co-ordinates (x, y, z) for the stationary system, and (ξ, η, ζ) for the moving system. Let the time t be determined for each point of the stationary system (which are provided with clocks) by means of the clocks which are placed in the stationary system, with the help of light-signals as described in § 1. Let also the time τ of the moving system be determined for each point of the moving system (in which there are clocks which are at rest relative to the moving system), by means of the method of light signals between these points (in which there are clocks) in the manner described in § 1.

To every value of (x, y, z, t) which fully determines the position and time of an event in the stationary system, there correspond a system of values (ξ, η, ζ, τ); now the problem is to find out the system of equations connecting these magnitudes.

Primarily it is clear that on account of the property of homogeneity which we ascribe to time and space, the equations must be linear.

If we put $x' = x - vt$, then it is clear that at a point relatively at rest in the system k, we have a system of values $(x' y z)$ which are independent of time. Now let us find out τ as a function of (x', y, z, t). For this purpose we have to express in equations the fact that τ is not other than the time given by the clocks which are at rest in the system k which must be made synchronous in the manner described in § 1.

Let a ray of light be sent at time τ_0 from the origin of the system k along the X-axis towards x' and let it be reflected from that place at time τ_1 towards the origin of moving co-ordinates and let it arrive there at time τ_2; then we must have

$$\tfrac{1}{2}(\tau_0 + \tau_2) = \tau_1$$

If we now introduce the condition that τ is a function of co-ordinates, and apply the principle of constancy of the velocity of light in the stationary system, we have

$$\frac{1}{2}\left\{\tau(0, 0, 0, t) + \tau\left(0, 0, 0, \left\{t + \frac{x'}{c-v} + \frac{x'}{c+v}\right\}\right)\right] = \tau\left(x', 0, 0, t + \frac{x'}{c-v}\right).$$

It is to be noticed that instead of the origin of co-ordinates, we could select some other point as the exit point for rays of light, and therefore the above equation holds for all values of $(x', y, z, t,)$.

A similar conception, being applied to the y- and z-axis gives us, when we take into consideration the fact that light when viewed from the stationary system, is always propagated along those axes with the velocity $\sqrt{(c^2 - v^2)}$, we have the questions

$$\frac{\partial}{\partial y} = 0, \quad \frac{\partial \tau}{\partial z} = 0.$$

From these equations it follows that τ is a linear function of x' and t. From equations (1) we obtain

$$\tau = a \left(t - \frac{vx'}{c^2 - v^2}\right)$$

where a is an unknown function of v.

With the help of these results it is easy to obtain the magnitudes (ξ, η, ζ) if we express by means of equations the fact that light, when measured in the moving system is always propagated with the constant velocity c (as the principle of constancy of light velocity in conjunction with the principle of relativity requires). For a time $\tau = 0$, if the ray is sent in the direction of increasing ξ, we have

$$\xi = c\tau, \text{ i.e. } \xi = a\,c\left(t - \frac{vx'}{c^2 - v^2}\right)$$

Now the ray of light moves relative to the origin of k with a velocity $c - v$, measured in the stationary system; therefore we have

$$\frac{x'}{c - v} = t$$

Substituting these values of t in the equation for ξ, we obtain

c^2

$\xi = a$ ------------ x'

$c^2 - v^2$

In an analogous manner, we obtain by considering the ray of light which moves along the *y*-axis,

vx'

$\eta = c\tau = a\, c(t$ - -----------)

$c^2 - v^2$

where

y

------------- $= t, x' = 0,$

$\sqrt{(c^2 - v^2)}$

Therefore

c

$\eta = a$ -------------$y,$

$\sqrt{(c^2 - v^2)}$

c

$\zeta = a$ --------------- $z.$

$\sqrt{(c^2 - v^2)}$

If for x', we substitute its value $x - tv$, we obtain

$v.c$

$\tau = \varphi(v).\ \beta\ (t$ - ------- ,

c^2

$\xi = \varphi(v).\ \beta\ (x - vt)$,

$\eta = \varphi(v)\, y$

$\zeta = \varphi(v)\, z,$

where

$$\beta = \frac{1}{\sqrt{1 - \frac{v^2}{c^2}}}$$

and

$\varphi(v) = ac / \sqrt{(c^2 - v^2)} = a / \beta$

is a function of v.

If we make no assumption about the initial position of the moving system and about the null-point of t, then an additive constant is to be added to the right hand side.

We have now to show, that every ray of light moves in the moving system with a velocity c (when measured in the moving system), in case, as we have actually assumed, c is also the velocity in the stationary system; for we have not as yet adduced any proof in support of the assumption that the principle of relativity is reconcilable with the principle of constant light-velocity.

At a time $\tau = t = 0$ let a spherical wave be sent out from the common origin of the two systems of co-ordinates, and let it spread with a velocity c in the system K. If (x, y, z), be a point reached by the wave, we have

$x^2 + y^2 + z^2 = c^2t^2$

with the aid of our transformation-equations, let us transform this equation, and we obtain by a simple calculation,

$\xi^2 + \eta^2 + \zeta^2 = c^2\tau^2$.

Therefore the wave is propagated in the moving system with the same velocity c, and as a spherical wave.[7] Therefore we show that the two principles are mutually reconcilable.

In the transformations we have got an undetermined function $\varphi(v)$, and we now proceed to find it out.

Let us introduce for this purpose a third co-ordinate system k', which is set in motion relative to the system k, the motion being parallel to the ξ-axis. Let the velocity of the origin be $(-v)$. At the time $t = 0$, all the initial co-ordinate points coincide, and for $t = x = y = z = 0$, the time t' of the system $k' = 0$. We shall say that $(x' y' z' t')$ are the co-ordinates measured in the system k', then by a two-fold application of the transformation-equations, we obtain

v

$\tau' = \varphi(-v)\beta(-v)\{\tau + --- \xi\}$

c^2

$= \varphi(v)\varphi(-v)t,$

$x' = \varphi](v)\beta(v)(\xi + v\tau)$

$= \varphi(v)\varphi(-v)x$, etc.

Since the relations between (x', y', z', t'), and (x, y, z, t) do not contain time explicitly, therefore K and k' are relatively at rest.

It appears that the systems K and k' are identical.

$\therefore \varphi(v)\varphi(-v) = 1.$

Let us now turn our attention to the part of the ξ-axis between ($\xi = 0, \eta = 0, \zeta = 0$), and ($\xi = 0, \eta = 1, \zeta = 0$). Let this piece of the y-axis be covered with a rod moving with the velocity v relative to the system K and perpendicular to its axis;—the ends of the rod having therefore the co-ordinates

$x_1 = vt, y_1 = l \,/\, \varphi(v), z_1 = 0$

$x_2 = vt, y_2 = 0, z_2 = 0$

Therefore the length of the rod measured in the system K is $l/\varphi(v)$. For the system moving with velocity $(-v)$, we have on grounds of symmetry,

$l \quad l$

$------ = ------$

$\varphi(v) \quad \varphi(-v)$

$\therefore \varphi(v) = \varphi(-v), \therefore \varphi(v) = 1.$

§ 4. The physical significance of the equations obtained concerning moving rigid bodies and moving clocks.

Let us consider a rigid sphere (*i.e.*, one having a spherical figure when tested in the stationary system) of radius R which is at rest relative to the system (K), and whose centre coincides with the origin of K then the equation of the surface of this sphere, which is moving with a velocity v relative to K, is

$\xi^2 + \eta^2 + \zeta^2 = R^2.$

At time $t = 0$, the equation is expressed by means of $(x, y, z, t,)$ as

$$\left(\frac{x^2}{\sqrt{1-\frac{v_2}{c^2}}}\right)^2 + y^2 + z^2 = R^2.$$

A rigid body which has the figure of a sphere when measured in the moving system, has therefore in the moving condition—when considered from the stationary system, the figure of a rotational ellipsoid with semi-axes

$$R\sqrt{1-\frac{v^2}{c^2}}, \quad R, \ R.$$

Therefore the y and z dimensions of the sphere (therefore of any figure also) do not appear to be modified by the motion, but the x dimension is shortened in the ratio

$$1 : \sqrt{1-\frac{v^2}{c^2}};$$

the shortening is the larger, the larger is *v*. For $v = c$, all moving bodies, when considered from a stationary system shrink into planes. For a velocity larger than the velocity of light, our propositions become meaningless; in our theory *c* plays the part of infinite velocity.

It is clear that similar results hold about stationary bodies in a stationary system when considered from a uniformly moving system.

Let us now consider that a clock which is lying at rest in the stationary system gives the time *t*, and lying at rest relative to the moving system is capable of giving the time τ; suppose it to be placed at the origin of the moving system *k*, and to be so arranged that it gives the time τ. How much does the clock gain, when viewed from the stationary system K? We have,

$$\tau = \frac{1}{\sqrt{1-\frac{v^2}{c^2}}}\left(t-\frac{v}{c^2}x\right), \text{ and } x=vt,$$

$$\therefore \tau - t = \left[1-\sqrt{1-\frac{v^2}{c^2}}\right] t.$$

Therefore the clock loses by an amount $\frac{1}{2}(v^2/c^2)$ per second of motion, to the second order of approximation.

From this, the following peculiar consequence follows. Suppose at two points A and B of the stationary system two clocks are given which are synchronous in the sense explained in § 3 when viewed from the stationary system. Suppose the clock at A to be set in motion in the line joining it with B, then after the arrival of the clock at B, they will no longer be found synchronous, but the clock which was set in motion from A will lag behind the clock which had been all along at B by an amount $\frac{1}{2}t(v^2/c^2)$, where t is the time required for the journey.

We see forthwith that the result holds also when the clock moves from A to B by a polygonal line, and also when A and B coincide.

If we assume that the result obtained for a polygonal line holds also for a curved line, we obtain the following law. If at A, there be two synchronous clocks, and if we set in motion one of them with a constant velocity along a closed curve till it comes back to A, the journey being completed in t-seconds, then after arrival, the last mentioned clock will be behind the stationary one by $\frac{1}{2}t(v^2/c^2)$ seconds. From this, we conclude that a clock placed at the equator must be slower by a very small amount than a similarly constructed clock which is placed at the pole, all other conditions being identical.

§ 5. Addition-Theorem of Velocities.

Let a point move in the system k (which moves with velocity v along the x-axis of the system K) according to the equation

$$\xi = w_\xi \tau, \quad \eta = w_\eta \tau, \quad \zeta = o,$$

where w_ξ and w_η are constants.

It is required to find out the motion of the point relative to the system K. If we now introduce the system of equations in § 3 in the equation of motion of the point, we obtain

$$x = \frac{w_\xi + v}{1 + \frac{v w_\xi}{c^2}} t, \quad y = \frac{\left(1 - \frac{v^2}{c^2}\right)^{\frac{1}{2}} w_\eta t}{1 + \frac{v w_\xi}{c^2}}, \quad z = o.$$

The law of parallelogram of velocities hold up to the first order of approximation. We can put

$$U^2 = \left(\frac{\partial x}{\partial t} \right)^2 + \left(\frac{\partial y}{\partial t} \right)^2, \quad w^2 = w_\xi^2 + w_\eta^2,$$

and

$$\alpha = \tan^{-1} \frac{w}{w_\xi}.$$

i.e., α is put equal to the angle between the velocities *v*, and *w*. Then we have—

$$U = \frac{\left[(v^2 + w^2 + 2\, vw \cos \alpha) - \left(\frac{vw \sin \alpha}{c} \right)^2 \right]^{\frac{1}{2}}}{1 + \frac{vw \cos \alpha}{c^2}}$$

It should be noticed that *v* and *w* enter into the expression for velocity symmetrically. If *w* has the direction of the ξ-axis of the moving system,

$$U = \frac{v+w}{1+\frac{vw}{c^2}}$$

From this equation, we see that by combining two velocities, each of which is smaller than *c*, we obtain a velocity which is always smaller than *c*. If we put $v = c - \chi$, and $w = c - \lambda$, where χ and λ are each smaller than *c*,[8]

$$U = c\frac{2c-\chi-\lambda}{2c-\chi-\lambda+\frac{\chi\lambda}{c^2}} < c.$$

It is also clear that the velocity of light *c* cannot be altered by adding to it a velocity smaller than *c*. For this case,

$$U = \frac{c+v}{1+\frac{cv}{c^2}} = c.$$

We have obtained the formula for U for the case when *v* and *w* have the same direction; it can also be obtained by combining two transformations according to section § 3. If in addition to the systems K, and k, we introduce the system k´, of which the initial point moves parallel to the ξ-axis with velocity *w*, then between the magnitudes, *x*, *y*, *z*, *t* and the corresponding magnitudes of k´, we obtain a system of equations, which differ from the equations in § 3, only in the respect that in place of v, we shall have to write,

$$(v+w)/\left(1+\frac{vw}{c^2}\right).$$

We see that such a parallel transformation forms a group.

We have deduced the kinematics corresponding to our two fundamental principles for the laws necessary for us, and we shall now pass over to their application in electrodynamics.

II.—ELECTRODYNAMICAL PART.

§ 6. Transformation of Maxwell's equations for Pure Vacuum.

On the nature of the Electromotive Force caused by motion in a magnetic field.

The Maxwell-Hertz equations for pure vacuum may hold for the stationary system K, so that

$$\frac{1}{c}\frac{\partial}{\partial t}[X, Y, Z] = \begin{vmatrix} \frac{\partial}{\partial x} & \frac{\partial}{\partial y} & \frac{\partial}{\partial z} \\ L & M & N \end{vmatrix}$$

and

$$\frac{1}{c}\frac{\partial}{\partial t}[L, M, N] = -\begin{vmatrix} \frac{\partial}{\partial x} & \frac{\partial}{\partial y} & \frac{\partial}{\partial z} \\ X & Y & Z \end{vmatrix}$$

"Equation 1."

where [X, Y, Z] are the components of the electric force, L, M, N are the components of the magnetic force.

If we apply the transformations in §3 to these equations, and if we refer the electromagnetic processes to the co-ordinate system moving with velocity v, we obtain,

$$\frac{1}{c}\frac{\partial}{\partial\tau}\left[X,\ \beta\left(Y-\frac{v}{c}N\right),\ \beta\left(Z+\frac{v}{c}M\right)\right]=$$

$$\begin{vmatrix} \frac{\partial}{\partial\xi} & \frac{\partial}{\partial\eta} & \frac{\partial}{\partial\zeta} \\ L & \beta\left(M+\frac{v}{c}Z\right) & \beta\left(N-\frac{v}{c}Y\right) \end{vmatrix}$$

and

$$\frac{1}{c}\frac{\partial}{\partial\tau}\left[L,\ \beta\left(M+\frac{v}{c}Z\right),\ \beta\left(N-\frac{v}{c}Y\right)\right]$$

$$=-\begin{vmatrix} \frac{\partial}{\partial\xi} & \frac{\partial}{\partial\eta} & \frac{\partial}{\partial\zeta} \\ X & \beta\left(Y-\frac{v}{c}N\right) & \beta\left(Z+\frac{v}{c}M\right) \end{vmatrix}, \ \ldots$$

where $\beta=\frac{1}{\sqrt{1-v^2/c^2}}$

"Equation 2."

where

$$\beta=\frac{1}{\sqrt{1-v^2/c^2}}$$

The principle of Relativity requires that the Maxwell-Hertzian equations for pure vacuum shall hold also for the system k, if they hold for the system K, *i.e.*, for the vectors of the electric and magnetic forces acting upon electric and magnetic masses in the moving system k, which are defined by their pondermotive reaction, the same equations hold, ... *i.e.* ...

$$\frac{1}{c}\frac{\partial}{\partial\tau}(X', Y', Z') = \begin{vmatrix} \frac{\partial}{\partial\xi} & \frac{\partial}{\partial\eta} & \frac{\partial}{\partial\zeta} \\ L' & M' & N' \end{vmatrix},$$

$$\frac{1}{c}\frac{\partial}{\partial\tau}(L', M', N') = -\begin{vmatrix} \frac{\partial}{\partial\xi} & \frac{\partial}{\partial\eta} & \frac{\partial}{\partial\zeta} \\ X' & Y' & Z' \end{vmatrix} \ldots$$

" Equation 3."

Clearly both the systems of equations (2) and (3) developed for the system k shall express the same things, for both of these systems are equivalent to the Maxwell-Hertzian equations for the system K. Since both the systems of equations (2) and (3) agree up to the symbols representing the vectors, it follows that the functions occurring at corresponding places will agree up to a certain factor $\psi(v)$, which depends only on v, and is independent of (ξ, η, ζ, τ). Hence the relations,

$$[X', Y', Z'] = \psi(v)\left[X, \beta\left(Y - \frac{v}{c}N\right), \beta\left(Z + \frac{v}{c}M\right)\right],$$

$$[L', M', N'] = \psi(v)\left[L, \beta\left(M - \frac{v}{c}Z\right), \beta\left(N + \frac{v}{c}Y\right)\right],$$

Then by reasoning similar to that followed in §(3), it can be shown that $\psi(v) = 1$.

$$[X', Y', Z'] = [X, \beta(Y - \frac{v}{c} N), \beta(Z + \frac{v}{c} M)]$$

$$[L', M', N'] = [L, \beta(M - \frac{v}{c} Z), \beta(N + \frac{v}{c} Y)],$$

For the interpretation of these equations, we make the following remarks. Let us have a point-mass of electricity which is of magnitude unity in the stationary system K, *i.e.*, it exerts a unit force upon a similar quantity placed at a distance of 1 cm. If this quantity of electricity be at rest in the stationary system, then the force acting upon it is equivalent to the vector (X, Y, Z) of electric force. But if the quantity of electricity be at rest relative to the moving system (at least for the moment considered), then the force acting upon it, and measured in the moving system is equivalent to the vector (X', Y', Z'). The first three of equations (1), (2), (3), can be expressed in the following way:—

1. If a point-mass of electric unit pole moves in an electro-magnetic field, then besides the electric force, an electromotive force acts upon it, which, neglecting the numbers involving the second and higher powers of v/c, is equivalent to the vector-product of the velocity vector, and the magnetic force divided by the velocity of light (Old mode of expression).

2. If a point-mass of electric unit pole moves in an electro-magnetic field, then the force acting upon it is equivalent to the electric force existing at the position of the unit pole, which we obtain by the transformation of the field to a co-ordinate system which is at rest relative to the electric unit pole [New mode of expression].

Similar theorems hold with reference to the magnetic force. We see that in the theory developed the electro-magnetic force plays the part of an auxiliary concept, which owes its introduction in theory to the circumstance that the electric and magnetic forces possess no existence independent of the nature of motion of the co-ordinate system.

It is further clear that the asymmetry mentioned in the introduction which occurs when we treat of the current excited by the relative motion of a magnet and a conductor disappears. Also the question about the seat of electromagnetic energy is seen to be without any meaning.

§ 7. Theory of Döppler's Principle and Aberration.

In the system K, at a great distance from the origin of co-ordinates, let there be a source of electrodynamic waves, which is represented with sufficient approximation in a part of space not containing the origin, by the equations:—

$$X = X_0 \sin \Phi$$

$$Y = Y_0 \sin \Phi$$

$$Z = Z_0 \sin \Phi$$

$$L = L_0 \sin \Phi$$

$$M = M_0 \sin \Phi$$

$$N = N_0 \sin \Phi$$

$$\Phi = \omega\left(t - \frac{lx + my + nz}{c}\right)$$

Here (X_0, Y_0, Z_0) and (L_0, M_0, N_0) are the vectors which determine the amplitudes of the train of waves, (l, m, n) are the direction-cosines of the wave-normal.

Let us now ask ourselves about the composition of these waves, when they are investigated by an observer at rest in a moving medium *k*:—By applying the equations of transformation obtained in §6 for the electric and magnetic forces, and the equations of transformation obtained in § 3 for the co-ordinates, and time, we obtain immediately:—

$$X' = X_0 \sin \Phi'$$

$$Y' = \beta\left(Y_0 - \frac{v}{c} N_0\right) \sin \Phi'$$

$$Z' = \beta\left(Z_0 - \frac{v}{c} M_0\right) \sin \Phi'$$

$$L' = L_0 \sin \Phi'$$

$$M' = \beta\left(M_0 - \frac{v}{c} Z_0\right) \sin \Phi'$$

$$N' = \beta\left(N_0 - \frac{v}{c} Y_0\right) \sin \Phi'$$

$$\Phi' = \omega'\left(t - \frac{l'\xi + m'\eta + n'\zeta}{c}\right)$$

where

$$\omega' = \omega\beta\left(1 - \frac{lv}{c}\right), \quad l' = \frac{l - \frac{v}{c}}{1 - \frac{lv}{c}}, \quad m' = \frac{m}{\beta\left(1 - \frac{lv}{c}\right)}, \quad n' = \frac{n}{\beta\left(1 - \frac{lv}{c}\right)}.$$

From the equation for ω' it follows:—If an observer moves with the velocity *v* relative to an infinitely distant source of light emitting waves of frequency ν, in such a manner that the line joining the source of light and the observer makes an angle of Φ with the velocity of the observer referred to a system of co-ordinates which is stationary with regard to the source, then the frequency ν' which is perceived by the observer is represented by the formula

$$\nu' = \nu \frac{1 - \cos\Phi \frac{v}{c}}{\sqrt{1 - \frac{v^2}{c^2}}}.$$

This is Döppler's principle for any velocity. If $\Phi = 0$, then the equation takes the simple form

$$\nu' = \nu \left(\frac{1 - \frac{v}{c}}{1 + \frac{v}{c}} \right)^{\frac{1}{2}}.$$

We see that—contrary to the usual conception—$\nu = \infty$, for $v = -c$.

If Φ' = angle between the wave-normal (direction of the ray) in the moving system, and the line of motion of the observer, the equation for l' takes the form

$$\cos \Phi' = \frac{\cos \Phi - \frac{v}{c}}{1 - \frac{v}{c} \cos \Phi}.$$

This equation expresses the law of observation in its most general form. If $\Phi = \pi/2$, the equation takes the simple form

v

cos Φ' = ---

c

We have still to investigate the amplitude of the waves, which occur in these equations. If A and A' be the amplitudes in the stationary and the moving systems (either electrical or magnetic), we have

$$A'^2 = A^2 \frac{\left(1 - \frac{v}{c} \cos \Phi\right)^2}{1 - \frac{v^2}{c^2}}.$$

If $\Phi = 0$, this reduces to the simple form

$$A'^2 = A^2 \frac{1 - \frac{v}{c}}{1 + \frac{v}{c}}.$$

From these equations, it appears that for an observer, which moves with the velocity c towards the source of light, the source should appear infinitely intense.

§ 8. Transformation of the Energy of the Rays of Light. Theory of the Radiation-pressure on a perfect mirror.

Since $A^2/8\pi$ is equal to the energy of light per unit volume, we have to regard $A^2/8\pi$ as the energy of light in the moving system. A'^2/A^2 would therefore denote the ratio between the energies of a definite light-complex "measured when moving" and "measured when stationary," the volumes of the light-complex measured in K and *k* being equal. Yet this is not the case. If *l*, *m*, *n* are the direction-cosines of the wave-normal of light in the stationary system, then no energy passes through the surface elements of the spherical surface

$$(x - clt)^2 + (y - cmt)^2 + (z - cnt)^2 = R^2,$$

which expands with the velocity of light. We can therefore say, that this surface always encloses the same light-complex. Let us now consider the quantity of energy, which this surface encloses, when regarded from the system *k*, *i.e.*, the energy of the light-complex relative to the system *k*.

Regarded from the moving system, the spherical surface becomes an ellipsoidal surface, having, at the time $\tau = 0$, the equation:—

$$\left(\beta\xi - l\beta\frac{v}{c}\xi\right)^2 + \left(\eta - m\beta\frac{v}{c}\xi\right)^2 + \left(\zeta - n\beta\frac{v}{c}\xi\right)^2 = R^2$$

If S = volume of the sphere, S' = volume of this ellipsoid, then a simple calculation shows that:

$$\frac{S'}{S} = \frac{\beta}{\sqrt{1 - \frac{v}{c}\cos\Phi}}$$

If E denotes the quantity of light energy measured in the stationary system, E' the quantity measured in the moving system, which are enclosed by the surfaces mentioned above, then

$$\frac{E'}{E} = \frac{\frac{A'^2}{8\pi} S'}{\frac{A^2}{8\pi} S} = \frac{1 - \frac{v}{c}\cos\Phi}{\sqrt{1 - v^2/c^2}}$$

If Φ = 0, we have the simple formula:—

$$\frac{E'}{E} = \left(\frac{1 - \frac{v}{c}}{1 + \frac{v}{c}} \right)^{\frac{1}{2}}$$

It is to be noticed that the energy and the frequency of a light-complex vary according to the same law with the state of motion of the observer.

Let there be a perfectly reflecting mirror at the co-ordinate-plane $\xi = 0$, from which the plane-wave considered in the last paragraph is reflected. Let us now ask ourselves about the light-pressure exerted on the reflecting surface and the direction, frequency, intensity of the light after reflexion.

Let the incident light be defined by the magnitudes A cos Φ, *v* (referred to the system K). Regarded from *k*, we have the corresponding magnitudes:

$$A' = A \frac{1 - \frac{v}{c}\cos\Phi}{\sqrt{1 - \frac{v^2}{c^2}}}$$

$$\cos\Phi' = \frac{\cos\Phi - \frac{v}{c}}{1 - \frac{v}{c}\cos\Phi}$$

$$\nu' = \nu \frac{1 - \frac{v}{c}\cos\Phi}{\sqrt{1 - \frac{v^2}{c^2}}}$$

For the reflected light we obtain, when the process is referred to the system *k*:—

A" = A', cos Φ" = -cos Φ", ν" = ν'

By means of a back-transformation to the stationary system, we obtain K, for the reflected light:—

$$\left.\begin{aligned}
A''' &= A'' \frac{1+\frac{v}{c}\cos\Phi''}{\sqrt{1-\frac{v^2}{c^2}}} = A\frac{1-2\frac{v}{c}\cos\Phi+\frac{v^2}{c^2}}{1-\frac{v^2}{c^2}}, \\
\cos\Phi''' &= \frac{\cos\Phi''+\frac{v}{c}}{1+\frac{v}{c}\cos\Phi''} = -\frac{\left(1+\frac{v^2}{c^2}\right)\cos\Phi-2\frac{v}{c}}{1-2\frac{v}{c}\cos\Phi+\frac{v^2}{c^2}}, \\
\nu''' &= \nu''\frac{1+\frac{v}{c}\cos\Phi''}{\sqrt{1-\frac{v^2}{c^2}}} = \nu\frac{1-2\frac{v}{c}\cos\phi+\frac{v^2}{c^2}}{\left(1-\frac{v}{c}\right)^2}.
\end{aligned}\right\}$$

The amount or energy falling upon the unit surface of the mirror per unit of time (measured in the stationary system) is $A^2/(8\pi\ (c\cos\Phi - v))$. The amount of energy going away from unit surface of the mirror per unit of time is $A'''^2/(8\pi\ (-c\cos\Phi'' + v))$. The difference of these two expressions is, according to the Energy principle, the amount of work exerted, by the pressure of light per unit of time. If we put this equal to $P.v$, where P = pressure of light, we have

$$P=2\frac{A^2}{8\pi}\frac{\left(\cos\Phi-\frac{v}{c}\right)^2}{1-\left(\frac{v}{c}\right)^2}$$

As a first approximation, we obtain

$$P = 2\,\frac{A^2}{8\pi}\cos^2\Phi$$

which is in accordance with facts, and with other theories.

All problems of optics of moving bodies can be solved after the method used here. The essential point is, that the electric and magnetic forces of light, which are influenced by a moving body, should be transformed to a system of co-ordinates which is stationary relative to the body. In this way, every problem of the optics of moving bodies would be reduced to a series of problems of the optics of stationary bodies.

§ 9. Transformation of the Maxwell-Hertz Equations.

Let us start from the equations:—

$$\left.\begin{aligned}
\frac{1}{c}\left(\rho u_x + \frac{\partial X}{\partial t}\right) &= \frac{\partial N}{\partial y} - \frac{\partial M}{\partial z} \\
\frac{1}{c}\left(\rho u_y + \frac{\partial Y}{\partial t}\right) &= \frac{\partial L}{\partial z} - \frac{\partial N}{\partial x} \\
\frac{1}{c}\left(\rho u_z + \frac{\partial Z}{\partial t}\right) &= \frac{\partial M}{\partial x} - \frac{\partial L}{\partial y}
\end{aligned}\right\}
\qquad
\left.\begin{aligned}
\frac{1}{c}\frac{\partial L}{\partial t} &= \frac{\partial Y}{\partial z} - \frac{\partial Z}{\partial y} \\
\frac{1}{c}\frac{\partial M}{\partial t} &= \frac{\partial Z}{\partial x} - \frac{\partial X}{\partial z} \\
\frac{1}{c}\frac{\partial N}{\partial t} &= \frac{\partial X}{\partial y} - \frac{\partial Y}{\partial x}
\end{aligned}\right\}$$

where

$$\rho = \frac{\partial X}{\partial x} + \frac{\partial Y}{\partial y} + \frac{Z\partial}{\partial z}$$

denotes 4π times the density of electricity, and (u_x, u_y, u_z) are the velocity-components of electricity. If we now suppose that the electrical-masses are bound unchangeably to small, rigid bodies (Ions, electrons), then these equations form the electromagnetic basis of Lorentz's electrodynamics and optics for moving bodies.

If these equations which hold in the system K, are transformed to the system *k* with the aid of the transformation-equations given in § 3 and § 6, then we obtain the equations:—

$$\frac{1}{c}\left[\rho' u_\xi + \frac{\partial X'}{\partial \tau}\right] = \frac{\partial N'}{\partial \eta} - \frac{\partial M'}{\partial \zeta}, \quad \frac{\partial L'}{\partial \tau} = \frac{\partial Y'}{\partial \zeta} - \frac{\partial Z'}{\partial \eta},$$

$$\frac{1}{c}\left[\rho' u_\eta + \frac{\partial Y'}{\partial \tau}\right] = \frac{\partial L'}{\partial \zeta} - \frac{\partial N'}{\partial \xi}, \quad \frac{\partial M'}{\partial \tau} = \frac{\partial Z'}{\partial \xi} - \frac{\partial X'}{\partial \zeta},$$

$$\frac{1}{c}\left[\rho' u_\zeta + \frac{\partial Z'}{\partial \tau}\right] = \frac{\partial M'}{\partial \xi} - \frac{\partial L'}{\partial \eta}, \quad \frac{\partial N'}{\partial \tau} = \frac{\partial X'}{\partial \eta} - \frac{\partial Y'}{\partial \xi},$$

where

$$\frac{u_x - v}{1 - \frac{u_x v}{c}} = u_\xi,$$

$$\frac{u_y}{\beta\left(1 - \frac{v u_x}{c^2}\right)} = u_\eta, \quad \rho' = \frac{\partial X'}{\partial \xi} + \frac{\partial Y'}{\partial \eta} + \frac{\partial Z'}{\partial \xi}$$

$$= \beta\left(1 - \frac{v u_x}{c^2}\right)\rho,$$

$$\frac{u_z}{\beta\left(1 - \frac{v u_x}{c^2}\right)} = u_\zeta,$$

Since the vector (u_ξ, u_η, u_ζ) is nothing but the velocity of the electrical mass measured in the system *k*, as can be easily seen from the addition-theorem of velocities in § 4—so it is hereby shown, that by taking our kinematical principle as the basis, the electromagnetic basis of Lorentz's theory of electrodynamics of moving bodies correspond to the relativity-postulate. It can be briefly remarked here that the following important law follows easily from the equations developed in the present section:—if an electrically charged body moves in any manner in space, and if its charge does not change thereby, when regarded from a system moving along with it, then the charge remains constant even when it is regarded from the stationary system K.

§ 10. Dynamics of the Electron (slowly accelerated).

Let us suppose that a point-shaped particle, having the electrical charge *e* (to be called henceforth the electron) moves in the electromagnetic field; we assume the following about its law of motion.

If the electron be at rest at any definite epoch, then in the next "*particle of time*," the motion takes place according to the equations

$$m\frac{d^2x}{dt^2} = eX,\quad m\frac{d^2y}{dt^2} = eY,\quad m\frac{d^2z}{dt^2} = eZ$$

Where (x, y, z) are the co-ordinates of the electron, and m is its mass.

Let the electron possess the velocity v at a certain epoch of time. Let us now investigate the laws according to which the electron will move in the 'particle of time' immediately following this epoch.

Without influencing the generality of treatment, we can and we will assume that, at the moment we are considering, the electron is at the origin of co-ordinates, and moves with the velocity v along the X-axis of the system. It is clear that at this moment ($t = 0$) the electron is at rest relative to the system *k*, which moves parallel to the X-axis with the constant velocity v.

From the suppositions made above, in combination with the principle of relativity, it is clear that regarded from the system *k*, the electron moves according to the equations

$$m\frac{d^2\xi}{d\tau^2} = eX',\quad m\frac{d^2\eta}{d\tau^2} = eY',\quad m\frac{d^2\zeta}{d\tau^2} = eZ',$$

in the time immediately following the moment, where the symbols (ξ, η, ζ, τ, X', Y', Z') refer to the system k. If we now fix, that for $t = v = y = z = 0$, $\tau = \xi = \eta = \zeta = 0$, then the equations of transformation given in § 3 (and § 6) hold, and we have:

$$\tau = \beta\left(t - \frac{v}{c^2}x\right), \quad \xi = \beta(x - vt), \quad \eta = y, \quad \zeta = z,$$

$$X' = X, \quad Y' = \beta\left(Y - \frac{v}{c}N\right), \quad Z' = \beta\left(Z + \frac{v}{c}M\right)$$

With the aid of these equations, we can transform the above equations of motion from the system k to the system K, and obtain:—

$$\left.\begin{aligned} \frac{d^2x}{dt^2} &= \frac{e}{m}\frac{1}{\beta^3}X, \quad \frac{d^2y}{dt^2} = \frac{e}{m}\frac{1}{\beta}\left(Y - \frac{v}{c}N\right), \\ \frac{d^2z}{dt^2} &= \frac{e}{m}\frac{1}{\beta}\left(Z + \frac{v}{c}M\right) \end{aligned}\right\}$$

Let us now consider, following the usual method of treatment, the longitudinal and transversal mass of a moving electron. We write the equations (A) in the form

$$m\beta^2 \frac{d^2x}{dt^2} = eX = eX'$$

$$m\beta^2 \frac{d^2y}{dt^2} = e\beta\left(Y - \frac{v}{c}N\right) = eY'$$

$$m\beta^2 \frac{d^2 z}{dt^2} = e\beta \left(Z - \frac{v}{c} M\right) = eZ'$$

and let us first remark, that eX', eY', eZ' are the components of the ponderomotive force acting upon the electron, and are considered in a moving system which, at this moment, moves with a velocity which is equal to that of the electron. This force can, for example, be measured by means of a spring-balance which is at rest in this last system. If we briefly call this force as "the force acting upon the electron," and maintain the equation:—

Mass-number × acceleration-number = force-number, and if we further fix that the accelerations are measured in the stationary system K, then from the above equations, we obtain:—

Longitudinal mass:

$$\frac{m}{\left(\sqrt{1-\frac{v^2}{c^2}}\right)^{\frac{3}{2}}}$$

Transversal mass:

$$\frac{m}{\sqrt{1-\frac{v^2}{c^2}}}$$

Naturally, when other definitions are given of the force and the acceleration, other numbers are obtained for the mass; hence we see that we must proceed very carefully in comparing the different theories of the motion of the electron.

We remark that this result about the mass hold also for ponderable material mass; for in our sense, a ponderable material point may be made into an electron by the addition of an electrical charge which may be as small as possible.

Let us now determine the kinetic energy of the electron. If the electron moves from the origin of co-ordinates of the system K with the initial velocity 0 steadily along the X-axis under the action of an electromotive force X, then it is clear that the energy drawn from the electrostatic field has the value $\int eXdx$. Since the electron is only slowly accelerated, and in consequence, no energy is given out in the form of radiation, therefore the energy drawn from the electro-static field may be put equal to the energy W of motion. Considering the whole process of motion in questions, the first of equations A) holds, we obtain:—

$$W = \int eXdx = \int_0^v m\beta^3 v\,dv = mc^2 \left[\frac{1}{\sqrt{1 - \frac{v^2}{c^2}}} - 1 \right]$$

For $v = c$, W is infinitely great. As our former result shows, velocities exceeding that of light can have no possibility of existence.

In consequence of the arguments mentioned above, this expression for kinetic energy must also hold for ponderable masses.

We can now enumerate the characteristics of the motion of the electrons available for experimental verification, which follow from equations A).

1. From the second of equations A), it follows that an electrical force Y, and a magnetic force N produce equal deflexions of an electron moving with the velocity v, when $Y = Nv/c$. Therefore we see that according to our theory, it is possible to obtain the velocity of an electron from the ratio of the magnetic deflexion A_m, and the electric deflexion A_e, by applying the law:—

$$\frac{A_m}{A_e} = \frac{v}{c}$$

This relation can be tested by means of experiments because the velocity of the electron can be directly measured by means of rapidly oscillating electric and magnetic fields.

2. From the value which is deduced for the kinetic energy of the electron, it follows that when the electron falls through a potential difference of P, the velocity *v* which is acquired is given by the following relation:—

$$P = \int X dx = \frac{m}{e} c^2 \left[\frac{1}{\sqrt{1 - \frac{v^2}{c^2}}} - 1 \right]$$

3. We calculate the radius of curvature R of the path, where the only deflecting force is a magnetic force N acting perpendicular to the velocity of projection. From the second of equations A) we obtain:

$$-\frac{d^2y}{dt^2} = \frac{v^2}{R} = \frac{e}{m} \frac{v}{c} N \sqrt{1 - \frac{v^2}{c^2}}$$

or

$$R = \frac{mv\beta c}{eN}$$

These three relations are complete expressions for the law of motion of the electron according to the above theory.

ALBRECHT EINSTEIN
[A short biographical note.]

The name of Prof. Albrecht Einstein has now spread far beyond the narrow pale of scientific investigators owing to the brilliant confirmation of his predicted deflection of light-rays by the gravitational field of the sun during the total solar eclipse of May 29, 1919. But to the serious student of science, he has been known from the beginning of the current century, and many dark problems in physics has been illuminated with the lustre of his genius, before, owing to the latest sensation just mentioned, he flashes out before public imagination as a scientific star of the first magnitude.

Einstein is a Swiss-German of Jewish extraction, and began his scientific career as a privat-dozent in the Swiss University of Zürich about the year 1902. Later on, he migrated to the German University of Prague in Bohemia as ausser-ordentliche (or associate) Professor. In 1914, through the exertions of Prof. M. Planck of the Berlin University, he was appointed a paid member of the Royal (now National) Prussian Academy of Sciences, on a salary of 18,000 marks per year. In this post, he has only to do and guide research work. Another distinguished occupant of the same post was Van't Hoff, the eminent physical chemist.

It is rather difficult to give a detailed, and consistent chronological account of his scientific activities,—they are so variegated, and cover such a wide field. The first work which gained him distinction was an investigation on Brownian Movement. An admirable account will be found in Perrin's book 'The Atoms.' Starting from Boltzmann's theorem connecting the entropy, and the probability of a state, he deduced a formula on the mean displacement of small particles (colloidal) suspended in a liquid. This formula gives us one of the best methods for finding out a very fundamental number in physics—namely—the number of molecules in one gm. molecule of gas (Avogadro's number). The formula was shortly afterwards verified by Perrin, Prof. of Chemical Physics in the Sorbonne, Paris.

To Einstein is also due the resuscitation of Planck's quantum theory of energy-emission. This theory has not yet caught the popular imagination to the same extent as the new theory of Time, and Space, but it is none the less iconoclastic in its scope as far as classical concepts are concerned. It was known for a long time that the observed emission of light from a heated black body did not correspond to the formula which could be deduced from the older classical theories of continuous emission and propagation. In the year 1900, Prof. Planck of the Berlin University worked out a formula which was based on the bold assumption that energy was emitted and absorbed by the molecules in multiples of the quantity $h\nu$, where h is a constant (which is universal like the constant of gravitation), and ν is the frequency of the light.

The conception was so radically different from all accepted theories that in spite of the great success of Planck's radiation formula in explaining the observed facts of black-body radiation, it did not meet with much favour from the physicists. In fact, some one remarked jocularly that according to Planck, energy flies out of a radiator like a swarm of gnats.

But Einstein found a support for the new-born concept in another direction. It was known that if green or ultraviolet light was allowed to fall on a plate of some alkali metal, the plate lost electrons. The electrons were emitted with all velocities, but there is generally a maximum limit. From the investigations of Lenard and Ladenburg, the curious discovery was made that this maximum velocity of emission did not at all depend upon the intensity of light, but upon its wavelength. The more violet was the light, the greater was the velocity of emission.

To account for this fact, Einstein made the bold assumption that the light is propagated in space as a unit pulse (he calls it a Light-cell), and falling upon an individual atom, liberates electrons according to the energy equation

$$h\nu = \frac{1}{2} mv^2 + A,$$

where (m, v) are the mass and velocity of the electron. A is a constant characteristic of the metal plate.

There was little material for the confirmation of this law when it was first proposed (1905), and eleven years elapsed before Prof. Millikan established, by a set of experiments scarcely rivalled for the ingenuity, skill, and care displayed, the absolute truth of the law. As results of this confirmation, and other brilliant triumphs, the quantum law is now regarded as a fundamental law of Energetics. In recent years, X-rays have been added to the domain of light, and in this direction also, Einstein's photo-electric formula has proved to be one of the most fruitful conceptions in Physics.

The quantum law was next extended by Einstein to the problems of decrease of specific heat at low temperature, and here also his theory was confirmed in a brilliant manner.

We pass over his other contributions to the equation of state, to the problems of null-point energy, and photo-chemical reactions. The recent experimental works of Nernst and Warburg seem to indicate that through Einstein's genius, we are probably for the first time having a satisfactory theory of photo-chemical action.

In 1915, Einstein made an excursion into Experimental Physics, and here also, in his characteristic way, he tackled one of the most fundamental concepts of Physics. It is well-known that according to Ampere, the magnetisation of iron and iron-like bodies, when placed within a coil carrying an electric current is due to the excitation in the metal of small electrical circuits. But the conception though a very fruitful one, long remained without a trace of experimental proof, though after the discovery of the electron, it was generally believed that these molecular currents may be due to the rotational motion of free electrons within the metal. It is easily seen that if in the process of magnetisation, a number of electrons be set into rotatory motion, then these will impart to the metal itself a turning couple. The experiment is a rather difficult one, and many physicists tried in vain to observe the effect. But in collaboration with de Haas, Einstein planned and successfully carried out this experiment, and proved the essential correctness of Ampere's views.

Einstein's studies on Relativity were commenced in the year 1905, and has been continued up to the present time. The first paper in the present collection forms Einstein's first great contribution to the Principle of Special Relativity. We have recounted in the introduction how out of the chaos and disorder into which the electrodynamics and optics of moving bodies had fallen previous to 1895, Lorentz, Einstein and Minkowski have succeeded in building up a consistent, and fruitful new theory of Time and Space.

But Einstein was not satisfied with the study of the special problem of Relativity for uniform motion, but tried, in a series of papers beginning from 1911, to extend it to the case of non-uniform motion. The last paper in the present collection is a translation of a comprehensive article which he contributed to the Annalen der Physik in 1916 on this subject, and gives, in his own words, the Principles of Generalized Relativity. The triumphs of this theory are now matters of public knowledge.

Einstein is now only 45, and it is to be hoped that science will continue to be enriched, for a long time to come, with further achievements of his genius.

Principle of Relativity

INTRODUCTION.

At the present time, different opinions are being held about the fundamental equations of Electro-dynamics for moving bodies. The Hertzian[9] forms must be given up, for it has appeared that they are contrary to many experimental results.

In 1895 H. A. Lorentz[10] published his theory of optical and electrical phenomena in moving bodies; this theory was based upon the atomistic conception (vorstellung) of electricity, and on account of its great success appears to have justified the bold hypotheses, by which it has been ushered into existence. In his theory, Lorentz proceeds from certain equations, which must hold at every point of "Äther"; then by forming the average values over "Physically infinitely small" regions, which however contain large numbers of electrons, the equations for electro-magnetic processes in moving bodies can be successfully built up.

In particular, Lorentz's theory gives a good account of the non-existence of relative motion of the earth and the luminiferous "Äther"; it shows that this fact is intimately connected with the covariance of the original equation, when certain simultaneous transformations of the space and time co-ordinates are effected; these transformations have therefore obtained from H. Poincare[11] the name of Lorentz-transformations. The covariance of these fundamental equations, when subjected to the Lorentz-transformation is a purely mathematical fact *i.e.* not based on any physical considerations; I will call this the Theorem of Relativity; this theorem rests essentially on the form of the differential equations for the propagation of waves with the velocity of light.

Now without *recognizing* any hypothesis about the connection between "Äther" and matter, we can expect these mathematically evident theorems to have their consequences so far extended—that thereby even those laws of ponderable media which are yet unknown may anyhow possess this covariance when subjected to a Lorentz-transformation; by saying this, we do not indeed express an opinion, but rather a conviction,—and this conviction I may be permitted to call the Postulate of Relativity. The position of affairs here is almost the same as when the Principle of Conservation of Energy was postulated in cases, where the corresponding forms of energy were unknown.

Now if hereafter, we succeed in maintaining this covariance as a definite connection between pure and simple observable phenomena in moving bodies, the definite connection may be styled 'the Principle of Relativity.'

These differentiations seem to me to be necessary for enabling us to characterise the present day position of the electro-dynamics for moving bodies.

H. A. Lorentz[12] has found out the "Relativity theorem" and has created the Relativity-postulate as a hypothesis that electrons and matter suffer contractions in consequence of their motion according to a certain law.

A. Einstein[13] has brought out the point very clearly, that this postulate is not an artificial hypothesis but is rather a new way of comprehending the time-concept which is forced upon us by observation of natural phenomena.

The Principle of Relativity has not yet been formulated for electro-dynamics of moving bodies in the sense characterized by me. In the present essay, while formulating this principle, I shall obtain the fundamental equations for moving bodies in a sense which is uniquely determined by this principle.

But it will be shown that none of the forms hitherto assumed for these equations can exactly fit in with this principle.[14]

We would at first expect that the fundamental equations which are assumed by Lorentz for moving bodies would correspond to the Relativity Principle. But it will be shown that this is not the case for the general equations which Lorentz has for any possible, and also for magnetic bodies; but this is approximately the case (if neglect the square of the velocity of matter in comparison to the velocity of light) for those equations which Lorentz hereafter infers for non-magnetic bodies. But this latter accordance with the Relativity Principle is due to the fact that the condition of non-magnetisation has been formulated in a way not corresponding to the Relativity Principle; therefore the accordance is due to the fortuitous compensation of two contradictions to the Relativity-Postulate. But meanwhile enunciation of the Principle in a rigid manner does not signify any contradiction to the hypotheses of Lorentz's molecular theory, but it shall become clear that the assumption of the contraction of the electron in Lorentz's theory must be introduced at an earlier stage than Lorentz has actually done.

In an appendix, I have gone into discussion of the position of Classical Mechanics with respect to the Relativity Postulate. Any easily perceivable modification of mechanics for satisfying the requirements of the Relativity theory would hardly afford any noticeable difference in observable processes; but would lead to very surprising consequences. By laying down the Relativity-Postulate from the outset, sufficient means have been created for deducing henceforth the complete series of Laws of Mechanics from the principle of conservation of Energy alone (the form of the Energy being given in explicit forms).

NOTATIONS.

Let a rectangular system ($x, y, z, t,$) of reference be given in space and time. The unit of time shall be chosen in such a manner with reference to the unit of length that the velocity of light in space becomes unity.

Although I would prefer not to change the notations used by Lorentz, it appears important to me to use a different selection of symbols, for thereby certain homogeneity will appear from the very beginning. I shall denote the vector electric force by E, the magnetic induction by M, the electric induction by e and the magnetic force by m, so that (E, M, e, m) are used instead of Lorentz's (E, B, D, H) respectively.

I shall further make use of complex magnitudes in a way which is not yet current in physical investigations, *i.e.*, instead of operating with (t), I shall operate with ($i\,t$), where i denotes $\sqrt{(-1)}$. If now instead of ($x, y, z, i\,t$), I use the method of writing with indices, certain essential circumstances will come into evidence; on this will be based a general use of the suffixes (1, 2, 3, 4). The advantage of this method will be, as I expressly emphasize here, that we shall have to handle symbols which have apparently a purely real appearance; we can however at any moment pass to real equations if it is understood that of the symbols with indices, such ones as have the suffix 4 only once, denote imaginary quantities, while those which have not at all the suffix 4, or have it twice denote real quantities.

An individual system of values of (x, y, z, t) *i. e.*, of ($x_1\ x_2\ x_3\ x_4$) shall be called a space-time point.

Further let u denote the velocity vector of matter, ε the dielectric constant, μ the magnetic permeability, σ the conductivity of matter, while ϱ denotes the density of electricity in space, and x the vector of "Electric Current" which we shall some across in §7 and §8.

PART I

§ 2.

The Limiting Case.

The Fundamental Equations for Äther.

By using the electron theory, Lorentz in his above mentioned essay traces the Laws of Electro-dynamics of Ponderable Bodies to still simpler laws. Let us now adhere to these simpler laws, whereby we require that for the limiting case $\varepsilon = 1$, $\mu = 1$, $\sigma = 0$, they should constitute the laws for ponderable bodies. In this ideal limiting case $\varepsilon = 1$, $\mu = 1$, $\sigma = 0$, E will be equal to *e*, and M to *m*. At every space time point (x, y, z, t) we shall have the equations[15]

(i) Curl $m - (\delta e/\delta t) = \varrho u$

(ii) div $e = \varrho$

(iii) Curl $e + \delta m/\delta t = 0$

(iv) div $m = 0$

I shall now write $(x_1\ x_2\ x_3\ x_4)$ for (x, y, z, t) and $(\varrho_1, \varrho_2, \varrho_3, \varrho_4)$ for

$$(\rho u_x, \rho u_y, \rho u_z, i\rho)$$

i.e. the components of the convection current ϱu, and the electric density multiplied by $\sqrt{-1}$

Further I shall write

$f_{23}, f_{31}, f_{12}, f_{14}, f_{24}, f_{34}.$

for

$m_x, m_y, m_z, -ie_x, -ie_y, -ie_z.$

i.e., the components of m and (-*i.e.*) along the three axes; now if we take any two indices (h. k) out of the series

3, 4), $f_{kh} = -f_{kh}$,

Therefore

$f_{32} = -f_{23}, f_{13} = -f_{31}, f_{21} = -f_{12}$

$f_{41} = -f_{14}, f_{44} = -f_{24}, f_{43} = -f_{34}$

Then the three equations comprised in (i), and the equation (ii) multiplied by i becomes

$$\left|\begin{array}{lllll} & \frac{\delta f_{12}}{\delta x_2} + & \frac{\delta f_{13}}{\delta x_3} + & \frac{\delta f_{14}}{\delta x_4} & = \rho_1 \\ \frac{\delta f_{21}}{\delta x_1} & + & \frac{\delta f_{23}}{\delta x_3} \times & \frac{\delta f_{24}}{\delta x_4} & = \rho_2 \\ \frac{\delta f_{31}}{\delta x_1} \times & \frac{\delta f_{32}}{\delta x_2} & + & \frac{\delta f_{34}}{\delta x_4} & = \rho_3 \\ \frac{\delta f_{41}}{\delta x_1} + & \frac{\delta f_{42}}{\delta x_2} + & \frac{\delta f_{43}}{\delta x_3} & & = \rho_4 \end{array}\right| \times$$

"Formula A."

On the other hand, the three equations comprised in (iii) and the (iv) equation multiplied by (*i*) becomes

$$\left|\begin{array}{lllll} & \frac{\delta f_{34}}{\delta x_2} + & \frac{\delta f_{42}}{\delta x_3} + & \frac{\delta f_{23}}{\delta x_4} & = 0 \\ \frac{\delta f_{43}}{\delta x_1} & + & \frac{\delta f_{14}}{\delta x_3} + & \frac{\delta f_{31}}{\delta x_4} & = 0 \\ \frac{\delta f_{24}}{\delta x_1} + & \frac{\delta f_{41}}{\delta x_2} & + & \frac{\delta f_{12}}{\delta x_4} & = 0 \\ \frac{\delta f_{32}}{\delta x_1} + & \frac{\delta f_{13}}{\delta x_2} + & \frac{\delta f_{21}}{\delta x_3} & & = 0 \end{array}\right| \times$$

"Formula B."

By means of this method of writing we at once notice the perfect symmetry of the 1st as well as the 2nd system of equations as regards permutation with the indices, (1, 2, 3, 4).

§ 3.

It is well-known that by writing the equations i) to iv) in the symbol of vector calculus, we at once set in evidence an invariance (or rather a (covariance) of the system of equations A) as well as of B), when the co-ordinate system is rotated through a certain amount round the null-point. For example, if we take a rotation of the axes round the z-axis, through an amount φ, keeping e, m fixed in space, and introduce new variables x_1' x_2' x_3' x_4' instead of x_1 x_2 x_3 x_4 where $x'_1 = x_1 \cos \varphi + x_2 \sin \varphi$, $x'_2 = -x_1 \sin \varphi + x_2 \cos \varphi$, $x'_3 = x_3$, $x'_4 = x_4$, and introduce magnitudes ϱ'_1, ϱ'_2, ϱ'_3, ϱ'_4, where $\varrho_1' = \varrho_1 \cos \varphi + \varrho_2 \sin \varphi$, $\varrho_2' = - \varrho_1 \sin \varphi + \varrho_2 \cos \varphi$ and $f'_{1\,2}, \ldots \ldots f'_{3\,4}$, where

$$f'_{23} = f_{23} \cos \varphi + f_{31} \sin \varphi,$$

$$f'_{31} = - f_{23} \sin \varphi + f_{31} \cos \varphi,$$

$$f'_{12} = f_{12},$$

$$f'_{14} = f_{14} \cos \varphi + f_{24} \sin \varphi,$$

$$f'_{24} = - f_{14} \sin \varphi + f_{24} \cos \varphi,$$

$$f'_{34} = f_{343\,4},$$

$$f'_{kh} = - f_{kh}\ (\text{h l k} = 1, 2, 3, 4).$$

then out of the equations (A) would follow a corresponding system of dashed equations (A´) composed of the newly introduced dashed magnitudes.

So upon the ground of symmetry alone of the equations (A) and (B) concerning the *suffixes* (1, 2, 3, 4), the theorem of Relativity, which was found out by Lorentz, follows without any calculation at all.

I will denote by $i\psi$ a purely imaginary magnitude, and consider the substitution

$$x_1' = x_1,$$

$$x_2' = x_2,$$

$$x_3' = x_3 \cos i\psi + x_4 \sin i\psi, \quad (1)$$

$$x_4'' = - x_3 \sin i\psi + x_4 \cos i\psi,$$

Putting

$$- i \tan i\psi = \frac{e^{\psi} - e^{-\psi}}{e^{\psi} + e^{-\psi}} = q, \; \psi = \frac{1}{2} \log \frac{1+q}{1-q}$$

"(2)."

We shall have $\cos i\psi = 1/\sqrt{(1 - q^2)}$, $\sin i\psi = iq/\sqrt{(1 - q^2)}$ where $-1 < q < 1$, and $\sqrt{(1 - q^2)}$ is always to be taken with the positive sign.

Let us now write $x'_1 = x'$, $x'_2 = y'$, $x'_3 = z'$, $x'_4 = it'$ (3)

then the substitution 1) takes the form

$x' = x, y' = y, z' = (z - qt)/\sqrt{(1 - q^2)}, t' = (-qz + t)/\sqrt{(1 - q^2)}$, (4)

the coefficients being essentially real.

If now in the above-mentioned rotation round the Z-axis, we replace 1, 2, 3, 4 throughout by 3, 4, 1, 2, and φ by $i\psi$, we at once perceive that simultaneously, new magnitudes $\varrho'_1, \varrho'_2, \varrho'_3, \varrho'_4$, where

$\varrho'_1 = \varrho_1, \varrho'_2 = \varrho_2, \varrho'_3 = \varrho_3 \cos i\psi + \varrho_4 \sin i\psi,$

$\varrho'_4 = - \varrho_3 \sin i\psi + \varrho_4 \cos i\psi),$

and $f'_{12} \ldots f'_{34}$, where

$f'_{41} = f_{41} \cos i\psi + f_{13} \sin i\psi,$

$f'_{13} = - f_{41} \sin i\psi + f_{13} \cos i\psi,$

$f'_{34} = f_{34},$

$f'_{32} = f_{32} \cos i\psi + f_{42} \sin i\psi,$

$f'_{42} = - f_{32} \sin i\psi + f_{42} \cos i\psi,$

$f'_{12} = f_{12}, f_{kh} = - f'_{kh},$

must be introduced. Then the systems of equations in (A) and (B) are transformed into equations (A´), and (B´), the new equations being obtained by simply dashing the old set.

All these equations can be written in purely real figures, and we can then formulate the last result as follows.

If the real transformations 4) are taken, and $x' y' z' t'$ be taken as a new frame of reference, then we shall have

(5) $\varrho' = \varrho [(-qu_z + 1)/\sqrt{(1 - q^2)}]$,

$\varrho' u_z' = \varrho[(u_z - q)/\sqrt{(1 - q^2)}]$,

$\varrho' u_x' = \varrho u_x$,

$\varrho' u_y' = \varrho u_y$.

(6) $e'_{x'} = (e_x - qm_y)/(\sqrt{(1 - q^2)})$,

$m'_{y'} = (qe_x + m_y)/(\sqrt{(1 - q^2)})$,

$e'_{z'} = e_z$.

(7) $m'_{x'} = (m_x - qe_y)/(\sqrt{(1 - q^2)})$,

$e'_{y'} = (qm_x + e_y)/(\sqrt{(1 - q^2)})$,

$m'_{z'} = m_z$.

Then we have for these newly introduced vectors u', e', m' (with components u_x', u_y', u_z'; e_x', e_y', e_z'; m_x', m_y', m_z'), and the quantity ϱ' a series of equations I'), II'), III'), IV') which are obtained from I), II), III), IV) by simply dashing the symbols.

We remark here that $e_x - qm_y$, $e_y + qm_x$ are components of the vector $e + [vm]$, where v is a vector in the direction of the positive Z-axis, and $| v | = q$, and $[vm]$ is the vector product of v and m; similarly $-qe_x + m_y$, $m_x + qe_y$ are the components of the vector $m - [ve]$.

The equations 6) and 7), as they stand in pairs, can be expressed as.

$e'_{x'} + im'_{x'} = (e_x + im_x) \cos i\psi + (e_y + im_y) \sin i\psi$,

$e'_{y'} + im'_{y'} = - (e_x + im_x) \sin i\psi + (e_y + im_y) \cos i\psi$,

$e'_{z'} + im'_{z'} = e'_z + im_z$.

If φ denotes any other real angle, we can form the following combinations:—

$(e'_{x'} + im'_{x'}) \cos. \varphi + (e'_{y''} + im'_{y'}) \sin \varphi$

$= (e_x + im_x) \cos. (\varphi + i\psi) + (e_y + im_y) \sin (\varphi + i\psi)$,

$$= (e'_{x'} + im'_{x'}) \sin \varphi + (e'_{y'} + im'_{y'}) \cos. \varphi$$

$$= - (e_x + im_x) \sin (\varphi + i\psi) + (e_y + im_y) \cos. (\varphi + i\psi).$$

§ 4. Special Lorentz Transformation.

The rôle which is played by the Z-axis in the transformation (4) can easily be transferred to any other axis when the system of axes are subjected to a transformation about this last axis. So we came to a more general law:—

Let v be a vector with the components v_x, v_y, v_z, and let $| v | = q < 1$. By $\square$ we shall denote any vector which is perpendicular to v, and by r_v, $r_\square$ we shall denote components of r in direction of $\square$ and v.

Instead of (x, y, z, t), new magnetudes $(x' y' z' t')$ will be introduced in the following way. If for the sake of shortness, r is written for the vector with the components (x, y, z) in the first system of reference, r' for the same vector with the components $(x' y' z')$ in the second system of reference, then for the direction of v, we have

(10) $r'_v = (r_v - qt)/\sqrt{(1 - q^2)}$

and for the perpendicular direction $\square$,

(11) $r'_\square = r_\square$

and further (12) $t' = (-qr_v + t)/\sqrt{(1 - q^2)}$.

The notations $(r'_\square, r'_v)$ are to be understood in the sense that with the directions v, and every direction $\square$ perpendicular to v in the system (x, y, z) are always associated the directions with the same direction cosines in the system $(x' y' z')$.

A transformation which is accomplished by means of (10), (11), (12) with the condition $0 < q < 1$ will be called a special Lorentz-transformation. We shall call v the vector, the direction of v the axis, and the magnitude of v the moment of this transformation.

If further ϱ' and the vectors u', e', m', in the system $(x' y' z')$ are so defined that,

(13) $\varrho' = \varrho[(-qu_v + 1)/\sqrt{(1 - q^2)}]$,

$\varrho' u'_v = \varrho(u_v - q)/\sqrt{(1 - q^2)}$,

$\varrho' u_\square = \varrho' u_v$,

further

(14) $(e' + im')_\square = ((e + im) - i[v, (e + im])']_\square)/\sqrt{(1 - q^2)}$.

$(15)\ (e' + im')_v = (e + im) - i[u, (e + im)]_v.$

Then it follows that the equations I), II), III), IV) are transformed into the corresponding system with dashes.

The solution of the equations (10), (11), (12) leads to

$$(16)\ r_v = (r'_v + qt')/\sqrt{(1 - q^2)},$$

$$r_{\square} = r'_{\square},$$

$$t = (qr'_v + t')/\sqrt{(1 - q^2)},$$

Now we shall make a very important observation about the vectors u and u'. We can again introduce the indices 1, 2, 3, 4, so that we write (x_1', x_2', x_3', x_4') instead of (x', y', z', it') and $\varrho_1', \varrho_2', \varrho_3', \varrho_4'$ instead of $(\varrho' u'\{x'\}, \varrho' u'\{y'\}, \varrho' u'\{z'\}, i\varrho')$.

Like the rotation round the Z-axis, the transformation (4), and more generally the transformations (10), (11), (12), are also linear transformations with the determinant + 1, so that

$$(17)\ x_1^2 + x_2^2 + x_3^2 + x_4^2 \text{ i. e. } x^2 + y^2 + z^2 - t^2,$$

is transformed into

$$x_1'^2 + x_2'^2 + x_3'^2 + x_4'^2 \text{ i. e. } x'^2 + y'^2 + z'^2 - t'^2.$$

On the basis of the equations (13), (14), we shall have $(\varrho_1^2 + \varrho_2^2 + \varrho_3^2 + \varrho_4^2) = \varrho^2(1 - u_{x^2}, -u_{y^2}, -u_{z^2}) = \varrho^2(1 - u^2)$ transformed into $\varrho^2(1 - u^2)$ or in other words,

$$(18)\ \varrho\sqrt{(1 - u^2)}$$

is an invariant in a Lorentz-transformation.

If we divide $(\varrho_1, \varrho_2, \varrho_3, \varrho_4)$ by this magnitude, we obtain the four values $(\omega_1, \omega_2, \omega_3, \omega_4) = (1/\sqrt{(1 - u^2)})(u_x, u_y, u_z, i)$ so that $\omega_1^2 + \omega_2^2 + \omega_3^2 + \omega_4^2 = -1$.

It is apparent that these four values are determined by the vector u and inversely the vector u of magnitude < 1 follows from the 4 values $\omega_1, \omega_2, \omega_3, \omega_4$; where $(\omega_1, \omega_2, \omega_3)$ are real, $-i\omega_4$ real and positive and condition (19) is fulfilled.

The meaning of $(\omega_1, \omega_2, \omega_3, \omega_4)$ here is, that they are the ratios of dx_1, dx_2, dx_3, dx_4 to

(20) $\sqrt{(-(dx_1^2 + dx_2^2 + dx_3^2 + dx_4^2))} = dt\sqrt{(1 - u^2)}$.

The differentials denoting the displacements of matter occupying the spacetime point (x_1, x_2, x_3, x_4) to the adjacent space-time point.

After the Lorentz-transformation is accomplished the velocity of matter in the new system of reference for the same space-time point $(x' y' z' t')$ is the vector u' with the ratios dx'/dt', dy'/dt', dz'/dt', dl'/dt', as components.

Now it is quite apparent that the system of values

$$x_1 = \omega_1, x_2 = \omega_2, x_3 = \omega_3, x_4 = \omega_4$$

is transformed into the values

$$x_1' = \omega_1', x_2' = \omega_2', x_3' = \omega_3', x_4' = \omega_4'$$

in virtue of the Lorentz-transformation (10), (11), (12).

The dashed system has got the same meaning for the velocity u' after the transformation as the first system of values has got for u before transformation.

If in particular the vector v of the special Lorentz-transformation be equal to the velocity vector u of matter at the space-time point (x_1, x_2, x_3, x_4) then it follows out of (10), (11), (12) that

$$\omega_1' = 0, \omega_2' = 0, \omega_3' = 0, \omega_4' = i$$

Under these circumstances therefore, the corresponding space-time point has the velocity $v' = 0$ after the transformation, it is as if we transform to rest. We may call the invariant $\rho\sqrt{(1 - u^2)}$ the rest-density of Electricity.[16]

§ 5. Space-time Vectors.
Of the 1st and 2nd kind.

If we take the principal result of the Lorentz transformation together with the fact that the system (A) as well as the system (B) is covariant with respect to a rotation of the coordinate-system round the null point, we obtain the general *relativity theorem.* In order to make the facts easily comprehensible, it may be more convenient to define a series of expressions, for the purpose of expressing the ideas in a concise form, while on the other hand I shall adhere to the practice of using complex magnitudes, in order to render certain symmetries quite evident.

Let us take a linear homogeneous transformation,

$$\begin{pmatrix} x_1 \\ x_2 \\ x_3 \\ x_4 \end{pmatrix} = \begin{pmatrix} a_{11} & a_{12} & a_{13} & a_{14} \\ a_{21} & a_{22} & a_{23} & a_{24} \\ a_{31} & a_{33} & a_{33} & a_{34} \\ a_{41} & a_{42} & a_{43} & a_{44} \end{pmatrix} \cdot \begin{pmatrix} x_1' \\ x_2' \\ x_3' \\ x_4' \end{pmatrix}$$

the Determinant of the matrix is +1, all co-efficients without the index 4 occurring once are real, while a_{41}, a_{42}, a_{43}, are purely imaginary, but a_{44} is real and > 0, and $x_1^2 + x_2^2 + x_3^2 + x_4^2$ transforms into $x_1'^2 + x_2'^2 + x_3'^2 + x_4'^2$. The operation shall be called a general Lorentz transformation.

(This notation, which is due to Dr. C. E. Cullis of the Calcutta University, has been used throughout instead of Minkowski's notation, $x_1 = a_{11}x_1' + a_{12}x_2' + a_{13}x_3' + a_{14}x_4'$.)

If we put $x_1' = x'$, $x_2' = y'$, $x_3' = z'$, $x_4' = it'$, then immediately there occurs a homogeneous linear transformation of (x, y, z, t) to (x', y', z', t') with essentially real co-efficients, whereby the aggregate $-x^2 - y^2 - z^2 + t^2$ transforms into $-x'^2 - y'^2 - z'^2 + t'^2$, and to every such system of values x, y, z, t with a positive t, for which this aggregate > 0, there always corresponds a positive t'; this last is quite evident from the continuity of the aggregate x, y, z, t.

The last vertical column of co-efficients has to fulfil the condition 22) $a_{14}^2 + a_{24}^2 + a_{34}^2 + a_{44}^2 = 1$.

If $a_{14} = a_{24} = a_{34} = 0$, then $a_{44} = 1$, and the Lorentz transformation reduces to a simple rotation of the spatial co-ordinate system round the world-point.

If a_{14}, a_{24}, a_{34} are not all zero, and if we put $a_{14} : a_{24} : a_{34} : a_{44} = v_x : v_y : v_z : i$

$$q = \sqrt{(v_x^2 + v_y^2 + v_z^2)} < 1.$$

On the other hand, with every set of values of $a_{14}, a_{24}, a_{34}, a_{44}$ which in this way fulfil the condition 22) with real values of v_x, v_y, v_z, we can construct the special Lorentz transformation (16) with $(a_{14}, a_{24}, a_{34}, a_{44})$ as the last vertical column,—and then every Lorentz-transformation with the same last vertical column $(a_{14}, a_{24}, a_{34}, a_{44})$ can be supposed to be composed of the special Lorentz-transformation, and a rotation of the spatial co-ordinate system round the null-point.

The totality of all Lorentz-Transformations forms a group. Under a space-time vector of the 1st kind shall be understood a system of four magnitudes $(\varrho_1, \varrho_2, \varrho_3, \varrho_4)$ with the condition that in case of a Lorentz-transformation it is to be replaced by the set $(\varrho_1', \varrho_2', \varrho_3', \varrho_4')$, where these are the values of (x_1', x_2', x_3', x_4'), obtained by substituting $(\varrho_1, \varrho_2, \varrho_3, \varrho_4)$ for (x_1, x_2, x_3, x_4) in the expression (21).

Besides the time-space vector of the 1st kind (x_1, x_2, x_3, x_4) we shall also make use of another space-time vector of the first kind (y_1, y_2, y_3, y_4), and let us form the linear combination

$$(23)\ f_{23}(x_2y_3 - x_3y_2) + f_{31}(x_3y_1 - x_1y_3) + f_{12}(x_1y_2 - x_2y_1) + f_{14}(x_1y_4 - x_4y_1) + f_{24}(x_2y_4 - x_4y_2) + f_{34}(x_3y_4 - x_4y_3)$$

with six coefficients f_{23}--f_{34}. Let us remark that in the vectorial method of writing, this can be constructed out of the four vectors.

$x_1, x_2, x_3; y_1, y_2, y_3; f_{23}, f_{31}, f_{12}; f_{14}, f_{24}, f_{34}$ and the constants x_4 and y_4, at the same time it is symmetrical with regard the indices (1, 2, 3, 4).

If we subject (x_1, x_2, x_3, x_4) and (y_1, y_2, y_3, y_4) simultaneously to the Lorentz transformation (21), the combination (23) is changed to:

$$(24)\ f_{23}'(x_2'y_3' - x_3'y_2') + f_{31}(x_3'y_1' - x_1'y_3') + f_{12}(x_1'y_2' - x_2'y_1') + f_{14}'(x_1'y_4') - x_4'y_1') + f_{24}'(x_2'y_4'$$

$- x_4'y_2') + f_{34}'(x_3'y_4' - x_4'y_3')$,

where the coefficients $f_{23}', f_{31}', f_{12}', f_{14}', f_{24}', f_{34}'$, depend solely on $(f_{23}\, f_{24})$ and the coefficients $a_{11} \ldots a_{44}$.

We shall define a space-time Vector of the 2nd kind as a system of six-magnitudes f_{23}, f_{31} ... f_{34}, with the condition that when subjected to a Lorentz transformation, it is changed to a new system f_{23}' ... f_{34}, ... which satisfies the connection between (23) and (24).

I enunciate in the following manner the general theorem of relativity corresponding to the equations (I)-(iv),—which are the fundamental equations for Äther.

If x, y, z, it (space co-ordinates, and time it) is subjected to a Lorentz transformation, and at the same time $(pu_x, pu_y, pu_z, i\varrho)$ (convection-current, and charge density ϱi) is transformed as a space time vector of the 1st kind, further $(m_x, m_y, m_z, -ie_x, -ie_y, -ie_z)$ (magnetic force, and electric induction $\times$ $(-i)$ is transformed as a space time vector of the 2nd kind, then the system of equations (I), (II), and the system of equations (III), (IV) transforms into essentially corresponding relations between the corresponding magnitudes newly introduced into the system.

These facts can be more concisely expressed in these words: the system of equations (I and II) as well as the system of equations (III) (IV) are covariant in all cases of Lorentz-transformation, where $(\varrho u, i\varrho)$ is to be transformed as a space time vector of the 1st kind, $(m - ie)$ is to be treated as a vector of the 2nd kind, or more significantly,—

$(\varrho u, i\varrho)$ is a space time vector of the 1st kind, $(m - ie)$[17] is a space-time vector of the 2nd kind.

I shall add a few more remarks here in order to elucidate the conception of space-time vector of the 2nd kind. Clearly, the following are invariants for such a vector when subjected to a group of Lorentz transformation.

$(i)\ m^2 - e^2 = f_{23}^2 + f_{31}^2 + f_{12}^2 + f_{14}^2 + f_{24}^2 + f_{24}^2$

$me = i(f_{23}f_{14} + f_{31}f_{24} + f_{12}f_{34})$.

A space-time vector of the second kind $(m - ie)$, where (m and e) are real magnitudes, may be called singular, when the scalar square $(m - ie)^2 = 0$, *ie* $m^2 - e^2 = 0$, and at the same time $(m\ e) = 0$, *ie* the vector m and e are equal and perpendicular to each other; when such is the case, these two properties remain conserved for the space-time vector of the 2nd kind in every Lorentz-transformation.

If the space-time vector of the 2nd kind is not singular, we rotate the spacial co-ordinate system in such a manner that the vector-product [*me*] coincides with the Z-axis, *i.e.* $m_x = 0$, $e_x = 0$. Then

$(m_x, -i\, e_x)^2 + (m_y, -i\, e_y)^2 \neq 0.$

Therefore $(e_y + i\, m_y)/(e_x + i\, e_x)$ is different from $+i$, and we can therefore define a complex argument $(\varphi + i\psi)$ in such a manner that

$$\tan(\varphi + i\psi) = \frac{e_y + i\, m_y}{e_x + i\, m_x}$$

If then, by referring back to equations (9), we carry out the transformation (1) through the angle ψ and a subsequent rotation round the Z-axis through the angle φ, we perform a Lorentz-transformation at the end of which $m_y = 0$, $e_y = 0$, and therefore m and e shall both coincide with the new Z-axis. Then by means of the invariants $m^2 - e^2$, (me) the final values of these vectors, whether they are of the same or of opposite directions, or whether one of them is equal to zero, would be at once settled.

§ 6. Concept of Time.

By the Lorentz transformation, we are allowed to effect certain *changes* of the time parameter. In consequence of this fact, it is no longer permissible to speak of the absolute simultaneity of two events. The ordinary idea of simultaneity rather presupposes that six independent parameters, which are evidently required for defining a system of space and time axes, are somehow reduced to three. Since we are accustomed to consider that these limitations represent in a unique way the actual facts very approximately, we maintain that the simultaneity of two events exists of themselves.[18] In fact, the following considerations will prove conclusive.

Let a reference system (x, y, z, t) for space time points (events) be somehow known. Now if a space point A (x_0, y_0, z_0) the time t_0 be compared with a space point P (x, y, z) at the time t, and if the difference of time $t - t_0$, (let $t > t_0$) be less than the length A P *i.e.* less than the time required for the propagation of light from A to P, and if $q = (t - t_0)/(\text{A P}) < 1$, then by a special Lorentz transformation, in which A P is taken as the axis, and which has the moment q, we can introduce a time parameter t', which (see equation 11, 12, § 4) has got the same value $t' = 0$ for both space-time points (A, t_0), and (P, t). So the two events can now be comprehended to be simultaneous.

Further, let us take at the same time $t_0 = 0$, two different space-points A, B, or three space-points (A, B, C) which are not in the same space-line, and compare therewith a space point P, which is outside the line A B, or the plane A B C, at another time t, and let the time difference $t - t_0$ ($t > t_0$) be less than the time which light requires for propagation from the line A B, or the plane (A B C) to P. Let q be the quotient of $(t - t_0)$ by the second time. Then if a Lorentz transformation is taken in which the perpendicular from P on A B, or from P on the plane A B C is the axis, and q is the moment, then all the three (or four) events (A, t_0), (B, t_0), (C, t_0) and (P, t) are simultaneous.

If four space-points, which do not lie in one plane, are conceived to be at the same time t_0, then it is no longer permissible to make a change of the time parameter by a Lorentz-transformation, without at the same time destroying the character of the simultaneity of these four space points.

To the mathematician, accustomed on the one hand to the methods of treatment of the poly-dimensional manifold, and on the other hand to the conceptual figures of the so-called non-Euclidean Geometry, there can be no difficulty in adopting this concept of time to the application of the Lorentz-transformation. The paper of Einstein which has been cited in the

Introduction, has succeeded to some extent in presenting the nature of the transformation from the physical standpoint.

PART II. ELECTRO-MAGNETIC PHENOMENA.

§ 7. Fundamental Equations for bodies at rest.

After these preparatory works, which have been first developed on account of the small amount of mathematics involved in the limiting case $\varepsilon = 1$, $\mu = 1$, $\sigma = 0$, let us turn to the electro-magnetic phenomena in matter. We look for those relations which make it possible for us—when proper fundamental data are given—to obtain the following quantities at every place and time, and therefore at every space-time point as functions of (x, y, z, t):—the vector of the electric force E, the magnetic induction M, the electrical induction e, the magnetic force m, the electrical space-density ϱ, the electric current s (whose relation hereafter to the conduction current is known by the manner in which conductivity occurs in the process), and lastly the vector v, the velocity of matter.

The relations in question can be divided into two classes.

Firstly—those equations, which,—when v, the velocity of matter is given as a function of (x, y, z, t),—lead us to a knowledge of other magnitude as functions of x, y, z, t—I shall call this first class of equations the fundamental equations—

Secondly, the expressions for the ponderomotive force, which, by the application of the Laws of Mechanics, gives us further information about the vector u as functions of (x, y, z, t).

For the case of bodies at rest, *i.e.* when $u\ (x, y, z, t) = 0$ the theories of Maxwell (Heaviside, Hertz) and Lorentz lead to the same fundamental equations. They are;—

(1) The Differential Equations:—which contain no constant referring to matter:—

(*i*) Curl $m - \delta e/\delta t = C$,

(*ii*) div $e = l\varrho$.

(*iii*) Curl E + $\delta M/\delta t = 0$,

(*iv*) Div M = 0.

(2) Further relations, which characterise the influence of existing matter for the most important case to which we limit ourselves *i.e.* for isotopic bodies;—they are comprised in the equations

(V) $e = \varepsilon E$, $M = \mu m$, $C = \sigma E$.

where ε = dielectric constant, μ = magnetic permeability, σ = the conductivity of matter, all given as function of x, y, z, t; s is here the conduction current.

By employing a modified form of writing, I shall now cause a latent symmetry in these equations to appear. I put, as in the previous work,

$x_1 = x, x_2 = y, x_3 = z, x_4 = it,$

and write s_1, s_2, s_3, s_4 for $C_x, C_y, C_z (\sqrt{-1})\varrho$.

Further $f_{23}, f_{31}, f_{12}, f_{14}, f_{24}, f_{34}$

for $m_x, m_y, m_z, -i(e_x, e_y, e_z)$,

and $F_{23}, F_{31}, F_{12}, F_{14}, F_{24}, F_{34}$

for $M_x, M_y, M_z, -i(E_x, E_y, E_z)$

lastly we shall have the relation $f_{kh} = -f_{hk}$, $F_{kh} = -F_{hk}$, (the letter f, F shall denote the field, s the (*i.e.* current).

Then the fundamental Equations can be written as

(A)

$$\partial f_{12}/\partial x_2 + \partial f_{13}/\partial x_3 + \partial f_{14}/\partial x_4 = s_1$$

$$\partial f_{21}/\partial x_1 + \quad + \partial f_{23}/\partial x_3 + \partial f_{24}/\partial x_4 = s_2$$

$$\partial f_{31}/\partial x_1 + \partial f_{32}/\partial x_2 + \quad + \partial f_{34}/\partial x_4 = s_3$$

$$\partial f_{41}/\partial x_1 + \partial f_{42}/\partial x_2 + \partial f_{43}/\partial x_3 \quad = s_4$$

and the equations (3) and (4), are

$$\partial F_{34}/\partial x_2 + \partial F_{42}/\partial x_3 + \partial F_{23}/\partial x_4 = 0$$

$$\partial F_{43}/\partial x_1 + \quad + \partial F_{14}/\partial x_3 + \partial F_{31}\partial x_4 = 0$$

$$\partial F_{24}/\partial x_1 + \partial F_{41}/\partial x_2 + \quad + \partial F_{12}/\partial x_4 = 0$$

$$\partial F_{32}/\partial x_1 + \partial F_{13}/\partial x_2 + \partial F_{21}/\partial x_3 \quad = 0$$

§ 8. The Fundamental Equations.

We are now in a position to establish in a unique way the fundamental equations for bodies moving in any manner by means of these three axioms exclusively.

The first Axion shall be,—

When a detached region[19] of matter is at rest at any moment, therefore the vector u is zero, for a system (x, y, z, t)—the neighbourhood may be supposed to be in motion in any possible manner, then for the space-time point x, y, z, t, the same relations (A) (B) (V) which hold in the case when all matter is at rest, shall also hold between ϱ, the vectors C, e, m, M, E and their differentials with respect to x, y, z, t. The second axiom shall be:—

Every velocity of matter is < 1, smaller than the velocity of propagation of light.[20]

The fundamental equations are of such a kind that when (x, y, z, it) are subjected to a Lorentz transformation and thereby (m - ie) and (M - iE) are transformed into space-time vectors of the second kind, (C, $i\varrho$) as a space-time vector of the 1st kind, the equations are transformed into essentially identical forms involving the transformed magnitudes.

Shortly I can signify the third axiom as:—

(m, -ie), and (M, -iE) are space-time vectors of the second kind, (C, ip) is a space-time vector of the first kind.

This axiom I call the Principle of Relativity.

In fact these three axioms lead us from the previously mentioned fundamental equations for bodies at rest to the equations for moving bodies in an unambiguous way.

According to the second axiom, the magnitude of the velocity vector $| u |$ is < 1 at any space-time point. In consequence, we can always write, instead of the vector u, the following set of four allied quantities

$$\omega_1 = u_x/\sqrt{(1 - u^2)},$$

$$\omega_2 = u_y/\sqrt{(1 - u^2)},$$

$$\omega_3 = u_z/\sqrt{(1 - u^2)},$$

$$\omega_4 = i/\sqrt{(1 - u^2)}$$

with the relation

$$(27)\ \omega_1^2 + \omega_2^2 + \omega_3^2 + \omega_4^2 = - |$$

From what has been said at the end of § 4, it is clear that in the case of a Lorentz-transformation, this set behaves like a space-time vector of the 1st kind.

Let us now fix our attention on a certain point (x, y, z) of matter at a certain time (t). If at this space-time point $u = 0$, then we have at once for this point the equations (A), (B) (V) of § 7. If $u \neq 0$, then there exists according to 16), in case $|u| < 1$, a special Lorentz-transformation, whose vector v is equal to this vector u (x, y, z, t), and we pass on to a new system of reference $(x' y' z' t')$ in accordance with this transformation. Therefore for the space-time point considered, there arises as in § 4, the new values 28) $\omega'_1 = 0$, $\omega'_2 = 0$, $\omega'_3 = 0$, $\omega'_4 = i$, therefore the new velocity vector $\omega' = 0$, the space-time point is as if transformed to rest. Now according to the third axiom the system of equations for the transformed point $(x' y' z' t)$ involves the newly introduced magnitude $(u'$ ρ', C', e', m', E', $M')$ and their differential quotients with respect to (x', y', z', t') in the same manner as the original equations for the point (x, y, z, t). But according to the first axiom, when $u' = 0$, these equations must be exactly equivalent to

(1) the differential equations (A'), (B'), which are obtained from the equations (A), (B) by simply dashing the symbols in (A) and (B).

(2) and the equations

(V') $e' = \varepsilon E'$, $M' = \mu m'$, $C' = \sigma E'$

where ε, μ, σ are the dielectric constant, magnetic permeability, and conductivity for the system $(x' y' z' t')$ *i.e.* in the space-time point $(x\, y, z\, t)$ of matter.

Now let us return, by means of the reciprocal Lorentz-transformation to the original variables (x, y, z, t), and the magnitudes (u, ρ, C, e, m, E, M) and the equations, which we then obtain from the last mentioned, will be the fundamental equations sought by us for the moving bodies.

Now from § 4, and § 6, it is to be seen that the equations $A)$, as well as the equations $B)$ are covariant for a Lorentz-transformation, *i.e.* the equations, which we obtain backwards from $A')$ $B')$, must be exactly of the same form as the equations $A)$ and $B)$, as we take them for bodies at rest. We have therefore as the first result:—

The differential equations expressing the fundamental equations of electrodynamics for moving bodies, when written in ρ and the vectors C, e, m, E, M, are exactly of the same form as the equations for moving bodies. The velocity of matter does not enter in these equations. In the vectorial way of writing, we have

I) curl $m - \partial e/\partial t = C_1$,

II) div $e = \varrho$

III) curl E + $\partial M/\partial t = 0$

IV) div M = 0

The velocity of matter occurs only in the auxiliary equations which characterise the influence of matter on the basis of their characteristic constants ε, μ, σ. Let us now transform these auxiliary equations V') into the original co-ordinates (*x*, *y*, *z*, and *t*.)

According to formula 15) in § 4, the component of e' in the direction of the vector u is the same as that of $(e + [u\ m])$, the component of m' is the same as that of $m - [u\ e]$, but for the perpendicular direction $\bar{u}$, the components of e', m' are the same as those of $(e + [u\ m])$ and $(m - [u\ e]$, multiplied by $1/\sqrt{(1 - u^2)}$. On the other hand E' and M' shall stand to E + [uM], and M - [uE] in the same relation as e' and m' to $e + [um]$, and $m - (ue)$. From the relation $e' = \varepsilon E'$, the following equations follow

(C) $e + [um] = \varepsilon(E + [uM])$,

and from the relation $M' = \mu m'$, we have

(D) $M - [u\ E] = \mu(m - [ue])$,

For the components in the directions perpendicular to u, and to each other, the equations are to be multiplied by $\sqrt{(1 - u^2)}$.

Then the following equations follow from the transformation? equations (12), (10), (11) in § 4, when we replace q, r_v, $r_{\square}$, t, r'_v, $r'_{\square}$, t' by $|u|$, C_u, $C_{\bar{u}}$, ϱ, C'_u, $C'_{\bar{u}}$, ϱ'

$\varrho' = (-|u|\ C_u + \varrho)/\sqrt{(1 - u^2)}$,

$C'_u = (C_u - |u|\varrho)/\sqrt{(1 - u^2)}$,

$C'_{\bar{u}} = C_{\bar{u}}$,

E) $(C_u - |u|\varrho)/\sqrt{(1 - u^2)} = \sigma(E + [uM])_u$,

$C_{\bar{u}} = \sigma\ (E + [uM])_u/\sqrt{(1 - u^2)}$.

In consideration of the manner in which σ enters into these relations, it will be convenient to call the vector $C - \varrho u$ with the components $C_u - \varrho|u|$ in the direction of u, and $C'_{\bar{u}}$ in the directions $\bar{u}$ perpendicular to u the "Convection current." This last vanishes for $\sigma = 0$.

We remark that for $\varepsilon = 1$, $\mu = 1$ the equations $e' = E'$, $m' = M'$ immediately lead to the equations $e = E$, $m = M$ by means of a reciprocal Lorentz-transformation with $-u$ as vector; and for $\sigma = 0$, the equation $C' = 0$ leads to $C = \varrho u$; that the fundamental equations of Äther discussed in § 2 becomes in fact the limitting case of the equations obtained here with $\varepsilon = 1$, $\mu = 1$, $\sigma = 0$.

§ 9. The Fundamental Equations in Lorentz's Theory.

Let us now see how far the fundamental equations assumed by Lorentz correspond to the Relativity postulate, as defined in §8. In the article on Electron-theory (Ency., Math., Wiss., Bd. V. 2, Art 14) Lorentz has given the fundamental equations for any possible, even magnetised bodies (see there page 209, Eqn XXX', formula (14) on page 78 of the same (part).

(III*a*") Curl (H - [*u*E]) = J + *d*D/*dt* + *u* div D

- curl [*u*D].

(I") div D = ϱ

(IV") curl E = - *d*B/*dt*, Div B = 0 (V')

Then for moving non-magnetised bodies, Lorentz puts (page 223, 3) μ = 1, B = H, and in addition to that takes account of the occurrence of the di-electric constant ε, and conductivity σ according to equations

(ε*q*XXXIV", p. 327) D - E = (ε - 1) {E + [*u*B]}

(ε*q*XXXIII', p. 223) J = σ(E + [*u*B])

Lorentz's E, D, H are here denoted by E, M, *e*, *m* while J denotes the conduction current.

The three last equations which have been just cited here coincide with eqn (II), (III), (IV), the first equation would be, if J is identified with C, = *u*ϱ (the current being zero for σ = 0,

(29) Curl [H - (*u*, E)] = C + *d*D/*dt* - curl [*u*D],

and thus comes out to be in a different form than (1) here. Therefore for magnetised bodies, Lorentz's equations do not correspond to the Relativity Principle.

On the other hand, the form corresponding to the relativity principle, for the condition of non-magnetisation is to be taken out of (D) in §8, with μ = 1, not as B = H, as Lorentz takes, but as (30) B - [*u*D] = H - [*u*D] (M - [*u*E] = *m* - [*ue*]. Now by putting H = B, the differential equation (29) is transformed into the same form as eqn (1) here when *m* - [*ue*] = M - [*u*E]. Therefore it so happens that by a compensation of two contradictions to the relativity principle, the differential equations of Lorentz for moving non-magnetised bodies at last agree with the relativity postulate.

If we make use of (30) for non-magnetic bodies, and put accordingly $H = B + [u, (D - E)]$, then in consequence of (C) in §8,

$(\varepsilon - 1) (E + [u, B]) = D - E + [u. [u, D - E]]$,

i.e. for the direction of u,

$(\varepsilon - 1) (E + [uB])_u = (D - E)_u$

and for a perpendicular direction ū,

$(\varepsilon - 1) [E + (uB)]_{\bar{u}} = (1 - u^2) (D - E)_{\bar{u}}$

i.e. it coincides with Lorentz's assumption, if we neglect u^2 in comparison to 1.

Also to the same order of approximation, Lorentz's form for J corresponds to the conditions imposed by the relativity principle [comp. (E) § 8]—that the components of J_u, $J_{\bar{u}}$ are equal to the components of $\sigma (E + [u B])$ multiplied by $\sqrt{(1 - u^2)}$ or $1 / \sqrt{(1 - u^2)}$ respectively.

§10. Fundamental Equations of E. Cohn.

E. Cohn assumes the following fundamental equations.

(31) Curl (M + [*u* E]) = *d*E/*dt* + u div. E + J

- Curl [E - (*u*. M)] = *d*M/*dt* + u div. M.

(32) J = σ E, = ε E - [*u* M], M = μ (*m* + [*u* E.])

where E M are the electric and magnetic field intensities (forces), E, M are the electric and magnetic polarisation (induction). The equations also permit the existence of true magnetism; if we do not take into account this consideration, div. M. is to be put = 0.

An objection to this system of equations, is that according to these, for ε = 1, μ = 1, the vectors force and induction do not coincide. If in the equations, we conceive E and M and not E - (U. M), and M + [U E] as electric and magnetic forces, and with a glance to this we substitute for E, M, E, M, div. E, the symbols *e*, M, E + [U M], *m* - [*u* *e*], ϱ, then the differential equations transform to our equations, and the conditions (32) transform into

J = σ(E + [*u* M])

e + [*u*, (*m* - [*u* *e*])] = ε(E + [*u* M])

M - [*u*, (E + *u* M)] = μ(*m* - [*u* *e*])

then in fact the equations of Cohn become the same as those required by the relativity principle, if errors of the order u^2 are neglected in comparison to 1.

It may be mentioned here that the equations of Hertz become the same as those of Cohn, if the auxiliary conditions are

(33) E = εE, M = μM, J = σE.

§11. Typical Representations of the Fundamental Equations.

In the statement of the fundamental equations, our leading idea had been that they should retain a covariance of form, when subjected to a group of Lorentz-transformations. Now we have to deal with ponderomotive reactions and energy in the electro-magnetic field. Here from the very first there can be no doubt that the settlement of this question is in some way connected with the simplest forms which can be given to the fundamental equations, satisfying the conditions of covariance. In order to arrive at such forms, I shall first of all put the fundamental equations in a typical form which brings out clearly their covariance in case of a Lorentz-transformation. Here I am using a method of calculation, which enables us to deal in a simple manner with the space-time vectors of the 1st, and 2nd kind, and of which the rules, as far as required are given below.

A system of magnitudes a_{hk} formed into the matrix

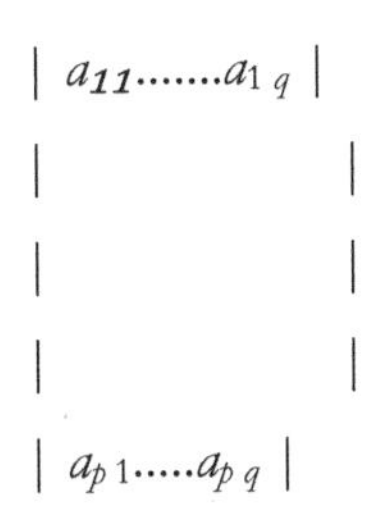

arranged in p horizontal rows, and q vertical columns is called a $p \times q$ series-matrix, and will be denoted by the letter A.

If all the quantities a_{hk} are multiplied by C, the resulting matrix will be denoted by CA.

If the roles of the horizontal rows and vertical columns be intercharged, we obtain a $q \times p$ series matrix, which will be known as the transposed matrix of A, and will be denoted by $\bar{A}$.

$\bar{A}$ = | a_{11} ...... a_{p1} |

| |

| a_{1q} ...... a_{pq} |

If we have a second $p \times q$ series matrix B,

B = | b_{11} .... b_{1q} |

| |

$| b_{p1} .... b_{pq} |$

then A + B shall denote the $p \times q$ series matrix whose members are $a_{hk} + b_{hk}$.

2^0 If we have two matrices

$A = | a_{11} .... a_{1q} |$

$| \quad |$

$| a_{p1} .... a_{pq} |$

$B = | b_{11} .... b_{1r} |$

$| \quad |$

$| b_{q1} .... b_{pr} |$

where the number of horizontal rows of B, is equal to the number of vertical columns of A, then by AB, the product of the matrices A and B, will be denoted the matrix

$C = | c_{11} .... c_{1r} |$

$| \quad |$

$| c_{pr} .... c_{pp} |$

where $c_{hk} = a_{h1} b_{1k} + a_{h2} b_{2h} + ... a_{ks} b_{sk} + ... + a_{kq} b_{qh}$

these elements being formed by combination of the horizontal rows of A with the vertical columns of B. For such a point, the associative law (AB)S = A(BS) holds, where S is a third matrix which has got as many horizontal rows as B (or AB) has got vertical columns.

For the transposed matrix of C = BA, we have $\dot{C} = \dot{B}\bar{A}$

3^0. We shall have principally to deal with matrices with at most four vertical columns and for horizontal rows.

As a unit matrix (in equations they will be known for the sake of shortness as the matrix 1) will be denoted the following matrix (4×4 series) with the elements.

(34) $| e_{11}\ e_{12}\ e_{13}\ e_{14} | = | 1\ 0\ 0\ 0 |$

$| e_{21}\ e_{22}\ e_{23}\ e_{24} | \quad | 0\ 1\ 0\ 0 |$

| e_{31} e_{32} e_{33} e_{34} | | 0 0 1 0 |

| e_{41} e_{42} e_{43} e_{44} | | 0 0 0 1 |

For a 4 × 4 series-matrix, Det A shall denote the determinant formed of the 4 × 4 elements of the matrix. If det A ≠ 0, then corresponding to A there is a reciprocal matrix, which we may denote by A^{-1} so that $A^{-1}A = 1$.

A matrix

$f =$ | 0 f_{12} f_{13} f_{14} |

| f_{21} 0 f_{23} f_{24} |

| f_{31} f_{32} 0 f_{34} |

| f_{41} f_{42} f_{43} 0 |

in which the elements fulfil the relation $f_{h\,k} = -f_{h\,k}$, is called an alternating matrix. These relations say that the transposed matrix □ = $-f$. Then by f^* will be the *dual*, alternating matrix

(35)

$f^* =$ | 0 f_{34} f_{42} f_{23} |

| f_{43} 0 f_{14} f_{31} |

| f_{24} f_{41} 0 f_{12} |

| f_{32} f_{13} f_{21} 0 |

Then (36) $f^* f = f_{34} f_{22} + f_{42} f_{31} + f_{32} f_{24}$

i.e. We shall have a 4 × 4 series matrix in which all the elements except those on the diagonal from left up to right down are zero, and the elements in this diagonal agree with each other, and are each equal to the above mentioned combination in (36).

The determinant of f is therefore the square of the combination, by $\mathrm{Det}^{1/2} f$ we shall denote the expression

$\mathrm{Det}^{1/2} f$

$= f_{32} f_{14} f_{13} f_{24} + f_{21} f_{34}.$

4^0. A linear transformation

$x_h = \alpha_{h1} x_1' + \alpha_{h2} x_2' + \alpha_{h3} x_3' + \alpha_{h4} x_4'$ (h = 1,2,3,

which is accomplished by the matrix

A = | $\alpha_{11}, \alpha_{12}, \alpha_{13}, \alpha_{14}$ |

| |

| $\alpha_{21}, \alpha_{22}, \alpha_{23}, \alpha_{24}$ |

| |

| $\alpha_{31}, \alpha_{32}, \alpha_{33}, \alpha_{34}$ |

| |

| $\alpha_{41}, \alpha_{42}, \alpha_{43}, \alpha_{44}$ |

will be denoted as the transformation A.

By the transformation A, the expression

$x^2_1 + x^2_2 + x^2_3 + x^2_4$ is changed into the quadratic for $m \sum \alpha_{hk} x_h' x_k'$,

where $\alpha_{hk} = \alpha_{1k} \alpha_{1k} + \alpha_{2h} \alpha_{2k} + \alpha_{3h} \alpha_{3k} + \alpha_{4h} \alpha_{4k}$ are the members of a 4 × 4 series matrix which is the product of Ā A, the transposed matrix of A into A. If by the transformation, the expression is changed to

$x'^2_1 + x_2'^2 + x_3'^2 + x'^2_4$,

we must have Ā A = 1.

A has to correspond to the following relation, if transformation (38) is to be a Lorentz-transformation. For the determinant of A) it follows out of (39) that (Det A)^2 = 1, or Det A = ± 1.

From the condition (39) we obtain

$A^{-1} = \bar{A}$,

i.e. the reciprocal matrix of A is equivalent to the transposed matrix of A.

For A as Lorentz transformation, we have further Det A = +1, the quantities involving the index 4 once in the subscript are purely imaginary, the other co-efficients are real, and $a_{44} > 0$.

5^0. A space time vector of the first kind[21] which s represented by the 1 × 4 series matrix,

(41) $s = |s_1 \; s_2 \; s_3 \; s_4|$

is to be replaced by sA in case of a Lorentz transformation

A. *i.e.* $s' = | s_1' s_2' s_3' s_4' | = | s_1 s_2 s_3 s_4 |$ A;

A space-time vector of the 2nd kind[22] with components $f_{23} \ldots f_{34}$ shall be represented by the alternating matrix

(42) $f = | 0 f_{12} f_{13} f_{14} |$

$| f_{21} 0 f_{23} f_{24} |$

$| f_{31} f_{32} 0 f_{34} |$

$| f_{41} f_{42} f_{43} 0 |$

and is to be replaced by $A^{-1} f A$ in case of a Lorentz transformation [see the rules in § 5 (23) (24)]. Therefore referring to the expression (37), we have the identity Det½ ($\bar{A} f A$) = Det A. Det½ f. Therefore Det½ f becomes an invariant in the case of a Lorentz transformation [see eq. (26) See. § 5].

Looking back to (36), we have for the dual matrix ($\bar{A} f^* A$) ($A^{-1} f A$) = $A^{-1} f^* f A$ = Det½ function. $A^{-1}A$ = Det½f from which it is to be seen that the dual matrix f^* behaves exactly like the primary matrix f, and is therefore a space time vector of the II kind; f^* is therefore known as the dual space-time vector of f with components (f_{14}, f_{24}, f_{34},), (f_{23}}, f_{31}, f_{12}).

6. If w and s are two space-time rectors of the 1st kind then by w □ (as well as by s □) will be understood the combination (43) $w_1 s_1 + w_2 s_2 + w_3 s_3 + w_4 s_4$.

In case of a Lorentz transformation A, since (wA) ($\bar{A}$□) = w s, this expression is invariant.—If w □ = 0, then w and s are perpendicular to each other.

Two space-time rectors of the first kind (w, s) gives us a 2 × 4 series matrix

$| w_1 w_2 w_3 w_4 |$

$| s_1 s_2 s_3 s_4 |$

Then it follows immediately that the system of six magnitudes (44)

$w_2 s_3 - w_3 s_2$,

$w_3 s_1 - w_1 s_3$,

$w_1 s_2 - w_2 s_1$,

$w_1 s_4 - w_4 s_1$,

$w_2 s_4 - w_4 s_2$,

$w_3 s_4 - w_4 s_3$,

behaves in case of a Lorentz-transformation as a space-time vector of the II kind. The vector of the second kind with the components (44) are denoted by $[w, s]$. We see easily that $\text{Det}^{1/2} [w, s] = 0$. The dual vector of $[w, s]$ shall be written as $[w, s]$.

If $\square$ is a space-time vector of the 1st kind, f of the second kind, $w f$ signifies a 1×4 series matrix. In case of a Lorentz-transformation A, w is changed into $w' = w\text{A}$, f into $f' = \text{A}^{-1} f \text{A}$,—therefore $w' f'$ becomes $= (w\text{A}\ \text{A}^{-1} f \text{A}) = w f \text{A}$ *i.e.* $w f$ is transformed as a space-time vector of the 1st kind.[23] We can verify, when w is a space-time vector of the 1st kind, f of the 2nd kind, the important identity

(45) $[w, wf] + [w, wf^*]^* = (w] \square)f.$

The sum of the two space time vectors of the second kind on the left side is to be understood in the sense of the addition of two alternating matrices.

For example, for $\omega_1 = 0, \omega_2 = 0, \omega_3 = 0, \omega_4 = i$,

$\omega f = |\ if_{41}, if_{42}, if_{43}, 0\ |$;

$\omega f^* = |\ if_{32}, if_{13}, if_{21}, 0\ |$

$[\omega \cdot \omega f] = 0, 0, 0, f_{41}, f_{42}, f_{43}$;

$[\omega \cdot \omega f^*]^* = 0, 0, 0, f_{32}, f_{13}, f_{21}$.

The fact that in this special case, the relation is satisfied, suffices to establish the theorem (45) generally, for this relation has a covariant character in case of a Lorentz transformation, and is homogeneous in $(\omega_1, \omega_2, \omega_3, \omega_4)$.

After these preparatory works let us engage ourselves with the equations (C,) (D,) (E) by means which the constants $\varepsilon\ \mu$, σ will be introduced.

Instead of the space vector u, the velocity of matter, we shall introduce the space-time vector of the first kind ω with the components.

$\omega_1 = u_x/\sqrt{(1 - u^2)}$,

$\omega_2 = u_y/\sqrt{(1 - u^2)}$,

$\omega_3 = u_z/\sqrt{(1 - u^2)}$,

$\omega_4 = i/\sqrt{(1 - u^2)}$.

(40) where $\omega_1^2 + \omega_2^2 + \omega_3^2 + \omega_4^2 = -1$ and $-i\omega_4 > 0$.

By F and f shall be understood the space time vectors of the second kind M - iE, m - ie.

In $\Phi = \omega F$, we have a space time vector of the first kind with components

$$\Phi_1 = \omega_2 F_{12} + \omega_3 F_{13} + \omega_4 F_{14}$$

$$\Phi_2 = \omega_1 F_{21} + \omega_3 F_{23} + \omega_4 F_{24}$$

$$\Phi_3 = \omega_1 F_{31} + \omega_2 F_{32} + \omega_4 F_{34}$$

$$\Phi_4 = \omega_1 F_{41} + \omega_2 F_{42} + \omega_3 F_{43}$$

The first three quantities (φ_1, φ_2, φ_3) are the components of the space-vector $(E + [u, M])/\sqrt{(1 - u^2)}$,

and further ($\varphi_4 = i[u\, E]/\sqrt{(1 - u^2)}$.

Because F is an alternating matrix,

(49) $\omega\Phi = \omega_1\, \varphi_1 + \omega_2\, \Phi_2 + \omega_3\, \Phi_3 + \omega_4\, \Phi_4 = 0$.

i.e. Φ is perpendicular to the vector ω; we can also write $\Phi_4 = i[\omega_x\, \Phi_1 + \omega_y\, \Phi_2 + \omega_z\, \Phi_3]$.

I shall call the space-time vector Φ of the first kind as the *Electric Rest Force.*[24]

Relations analogous to those holding between $-\omega F$, E, M, U, hold amongst $-\omega f$, e, m, u, and in particular $-\omega f$ is normal to ω. The relation (C) can be written as

{C} $\omega f = \varepsilon\omega F$.

The expression (ωf) gives four components, but the fourth can be derived from the first three.

Let us now form the time-space vector 1st kind, ψ - $i\omega f^*$, whose components are

$\psi_1 = -i(\omega_2 f_{34} + \omega_3 f_{42} + \omega_4 f_{23})$

$\psi_2 = -i(\omega_1 f_{43} + \omega_3 f_{44} + \omega_4 f_{31})$

$\psi_3 = -i(\omega_1 f_{24} + \omega_2 f_{41} + \omega_4 f_{12})$

$\psi_4 = -i(\omega_1 f_{32} + \omega_2 f_{13} + \omega_3 f_{21})$

Of these, the first three ψ_1, ψ_2, ψ_3, are the x, y, z components of the space-vector 51) $(m - (ue))/\sqrt{(1 - u^2)}$ and further (52) $\psi_4 = i(um)/\sqrt{(1 - u^2)}$.

Among these there is the relation

(53) $\omega\psi = \omega_1 \psi_1 + \omega_2 \psi_2 + \omega_3 \psi_3 + \omega_4 \psi_4 = 0$

which can also be written as $\psi_4 = i\,(u_x \psi_1 + u_y \psi_2 + u_z \psi_3)$.

The vector ψ is perpendicular to ω; we can call it the *Magnetic rest-force.*

Relations analogous to these hold among the quantities ωF^*, M, E, u and Relation (D) can be replaced by the formula

{ D } $-\omega F^* = \mu\psi f^*$.

We can use the relations (C) and (D) to calculate F and f from Φ and ψ we have

$\omega F = -\Phi$, $\omega F^* = -i\mu\psi$, $\omega f = -\varepsilon\Phi$, $\omega f^* = -i\psi$.

and applying the relation (45) and (46), we have

$F = [\omega.\ \Phi] + i\mu[\omega.\ \psi]^*$ 55)

$f = \varepsilon[\omega.\ \Phi] + i[\omega.\ \psi]^*$ 56)

i.e.

$F_{12} = (\omega_1 \Phi_1 - \omega_2 \Phi_1) + i\mu\ [\omega_3 \Psi_4 - \omega_4 \psi_3]$, etc.

$f_{12} = \varepsilon(\omega_1 \Phi_2 - \omega_2 \varphi_1) + i\ [\omega_3 \psi_4 - \omega_4 \psi_3]$., etc.

Let us now consider the space-time vector of the second kind $[\Phi\ \psi]$, with the components

$[\ \Phi_2 \psi_3 - \Phi_3 \psi_2, \Phi_3 \psi_1 - \Phi_1 \psi_3, \Phi_1 \psi_2 - \Phi_2 \psi_1\ ]$

$[\ \Phi_1 \psi_4 - \Phi_4 \psi_1, \Phi_2 \psi_4 - \Phi_4 \psi_2, \Phi_3 \psi_4 - \Phi_4 \psi_3\ ]$

Then the corresponding space-time vector of the first kind $\omega[\Phi, \psi]$ vanishes identically owing to equations 9) and 53)

for $\omega[\Phi.\psi] = -(\omega\psi)\Phi + (\omega\Phi)\psi$

Let us now take the vector of the 1st kind

(57) $\Omega = i\omega[\Phi\psi]^*$

with the components

$\Omega_1 = -i$ | ω_2 ω_3 ω_4 |

| Φ_2 Φ_3 Φ_4 |

| ψ_2 ψ_3 ψ_4 |, etc.

Then by applying rule (45), we have

(58) $[\Phi.\psi] = i[\omega\Omega]^*$

i.e. $\Phi_1\psi_2 - \Phi_2\psi_1 = i(\omega_3\Omega_4 - \omega_4\Omega_3)$ etc.

The vector Ω fulfils the relation

$(\omega\Omega) = \omega_1\Omega_1 + \omega_2\Omega_2 + \omega_3\Omega_3 + \omega_4\Omega_4 = 0,$

(which we can write as $\Omega_4 = i(\omega_x\Omega_1 + \omega_y\Omega_2 + \omega_z\Omega_3)$ and Ω is also normal to ω. In case $\omega = 0$, we have $\Phi_4 = 0$, $\psi_4 = 0$, $\Omega_4 = 0$, and

$[\Omega_1, \Omega_2, \Omega_3 =$ | Φ_1 Φ_2 Φ_3 |

| ψ_1 ψ_2 ψ_3 |.

I shall call Ω, which is a space-time vector 1st kind the Rest-Ray.

As for the relation E), which introduces the conductivity σ we have $-\omega S = -(\omega_1 s_1 + \omega_2 s_2 + \omega_3 s_3 + \omega_4 s_4) = (-|u| C_u + \varrho)/\sqrt{(1 - u^2)} = \varrho'$.

This expression gives us the rest-density of electricity (see §8 and §4).

Then 61) $= s + (\omega\square)\omega$ represents a space-time vector of the 1st kind, which since $\omega\omega = -1$, is normal to ω, and which I may call the rest-current. Let us now conceive of the first three component of this vector as the (x-y-z) co-ordinates of the space-vector, then the component in the direction of u is

$$C_u - (|u|\varrho')/\sqrt{(1 - u^2)}$$
$$= (c_u - |u|\varrho)/\sqrt{(1 - u^2)}$$
$$= J_u/(1 - u^2)$$

and the component in a perpendicular direction is $C_{\bar{u}} = J_{\bar{u}}$.

This space-vector is connected with the space-vector $J = C - \varrho u$, which we denoted in §8 as the conduction-current.

Now by comparing with $\Phi = -\omega F$, the relation (E) can be brought into the form

$$\{E\}\ s + (\omega \Box)\omega = -\sigma\omega F,$$

This formula contains four equations, of which the fourth follows from the first three, since this is a space-time vector which is perpendicular to ω.

Lastly, we shall transform the differential equations (A) and (B) into a typical form.

§12. The Differential Operator Lor.

A 4×4 series matrix 62) $S = | S_{11}\ S_{12}\ S_{13}\ S_{14} | = | S_{kh} |$

$| S_{21}\ S_{22}\ S_{23}\ S_{24} |$

$| S_{31}\ S_{32}\ S_{33}\ S_{34} |$

$| S_{41}\ S_{42}\ S_{43}\ S_{44} |$

with the condition that in case of a Lorentz transformation it is to be replaced by $\bar{A}SA$, may be called a space-time matrix of the II kind. We have examples of this in:—

1) the alternating matrix f, which corresponds to the space-time vector of the II kind,—

2) the product fF of two such matrices, for by a transformation A, it is replaced by $(A^{-1}fA \cdot A^{-1}FA) = A^{-1}fFA$,

3) further when $(\omega_1, \omega_2, \omega_3, \omega_4)$ and $(\Omega_1, \Omega_2, \Omega_3, \Omega_4)$ are two space-time vectors of the 1st kind, the 4×4 matrix with the element $S_{hk} = \omega_h \Omega_k$,

lastly in a multiple L of the unit matrix of 4×4 series in which all the elements in the principal diagonal are equal to L, and the rest are zero.

We shall have to do constantly with functions of the space-time point (x, y, z, it), and we may with advantage

employ the 1×4 series matrix, formed of differential symbols,—

$| \partial/\partial x, \partial/\partial y, \partial/\partial z, \partial/i\partial t, |$

or (63) $| \partial/\partial x_1\ \partial/\partial x_2\ \partial/\partial x_3\ \partial/\partial x_4 |$

For this matrix I shall use the shortened from "lor."[25]

Then if S is, as in (62), a space-time matrix of the II kind, by lor S' will be understood the 1×4 series matrix

$| K_1\ K_2\ K_3\ K_4 |$

where $K_k = \partial S_{1k}/\partial x_1 + \partial S_{2k}/\partial x_2 + \partial S_{3k}/\partial x_3 + \partial S_{4h}/\partial x_4$.

When by a Lorentz transformation A, a new reference system $(x'_1\ x'_2\ x'_3\ x_4)$ is introduced, we can use the operator

$\text{lor}' = | \partial/\partial x_1'\ \partial/\partial x_2'\ \partial/\partial x_3'\ \partial/\partial x_4' |$

Then S is transformed to $S' = \bar{A}\,S\,A = |\,S'_{hk}\,|$, so by lor 'S' is meant the 1×4 series matrix, whose element are

$$K'_k = \partial S'_{1k}/\partial x_1' + \partial S'_{2k}/\partial x_2'$$

$$+ \partial S'_{3k}/\partial x_3' + \partial S'_{4k}/\partial x_4'.$$

Now for the differentiation of any function of $(x\ y\ z\ t)$ we have the rule $\partial/\partial x_k' = \partial/\partial x_1\ \partial x_1/\partial x_k' + \partial/\partial x_2\ \partial x_2/\partial x_k' + \partial/\partial x_3\ \partial x_3/\partial x_k' + \partial/\partial x_4\ \partial x_4/\partial x_k' = \partial/\partial x_1\ a_{1k} + \partial/\partial x_2\ a_{2k} + \partial/\partial x_3\ a_{3k} + \partial/\partial x_4\ a_{4k}$.

so that, we have symbolically lor' = lor A.

Therefore it follows that

lor 'S' = lor $(A\,A^{-1}\,SA)$ = (lor S)A.

i.e., lor S behaves like a space-time vector of the first kind.

If L is a multiple of the unit matrix, then by lor L will be denoted the matrix with the elements

$$|\ \partial L/\partial x_1\ \partial L/\partial x_2\ \partial L/\partial x_3\ \partial L/\partial x_4\ |$$

If s is a space-time vector of the 1st kind, then

$$\text{lor}\ \square = \partial s_1/\partial x_1 + \partial s_2/\partial x_2 + \partial s_3/\partial x_3 + \partial s_4/\partial x_4.$$

In case of a Lorentz transformation A, we have

$$\text{lor}\ '\square' = \text{lor}\ A.\ \bar{A}s = \text{lor}\ s.$$

i.e., lor s is an invariant in a Lorentz-transformation.

In all these operations the operator lor plays the part of a space-time vector of the first kind.

If f represents a space-time vector of the second kind,—lor f denotes a space-time vector of the first kind with the components

$$\partial f_{12}/\partial x_2 + \partial f_{13}/\partial x_3 + \partial f_{14}/\partial x_4,$$

$$\partial f_{21}/\partial x_1 + \partial f_{23}/\partial x_3 + \partial f_{24}/\partial x_4,$$

$$\partial f_{31}/\partial x_1 + \partial f_{32}/\partial x_2 + \partial f_{34}/\partial x_4,$$

$$\partial f_{41}/\partial x_1 + \partial f_{42}/\partial x_2 + \partial f_{43}/\partial x_3$$

So the system of differential equations (A) can be expressed in the concise form

{A} lor f = -*s*,

and the system (B) can be expressed in the form

{B} log F* = 0.

Referring back to the definition (67) for log □, we find that the combinations lor ([=(lor *f*)=]), and lor ([=(lor F*)]) vanish identically, when *f* and F* are alternating matrices. Accordingly it follows out of {A}, that

(68) $(\partial s_1/\partial x_1) + (\partial s_2/\partial x_2) + (\partial s_3/\partial x_3) + (\partial s_4/\partial x_4) = 0$,

while the relation

(69) lor (lor F*) = 0,

signifies that of the four equations in {B}, only three represent independent conditions.

I shall now collect the results.

Let ω denote the space-time vector of the first kind

$(u/\sqrt{(1 - u^2\}}), i/\sqrt{(1 - u^2)})$

(u = velocity of matter),

F the space-time vector of the second kind (M,-*i*E)

(M = magnetic induction, E = Electric force,

f the space-time vector of the second kind (*m*,-*ie*)

(*m* = magnetic force, *e* = Electric Induction.

s the space-time vector of the first kind (C, *i*ϱ)

(ϱ = electrical space-density, C - ϱ*u* = conductivity current,

ε = dielectric constant, μ = magnetic permeability,

σ = conductivity,

then the fundamental equations for electromagnetic processes in moving bodies are[26]

{A} lor f = -s

{B} log F* = 0

{C} $\omega f = \varepsilon\omega F$

{D} $\omega F^* = \mu\omega f^*$

{E} s + (ω☐), w = - σωF.

$\omega\,\tilde{\omega}$ = -1, and ωF, ωf, ωF*, ωf*, s + (ωs)ω which are space-time vectors of the first kind are all normal to ω, and for the system {B}, we have

lor (lor F*) = 0.

Bearing in mind this last relation, we see that we have as many independent equations at our disposal as are necessary for determining the motion of matter as well as the vector u as a function of x, y, z, t, when proper fundamental data are given.

§ 13. The Product of the Field-vectors fF.

Finally let us enquire about the laws which lead to the determination of the vector ω as a function of $(x, y, z, t.)$ In these investigations, the expressions which are obtained by the multiplication of two alternating matrices

$$f = \begin{vmatrix} 0 & f_{12} & f_{13} & f_{14} \\ f_{21} & 0 & f_{23} & f_{24} \\ f_{31} & f_{32} & 0 & f_{34} \\ f_{41} & f_{42} & f_{43} & 0 \end{vmatrix}$$

$$F = \begin{vmatrix} 0 & F_{12} & F_{13} & F_{14} \\ F_{21} & 0 & F_{23} & F_{24} \\ F_{31} & F_{32} & 0 & F_{34} \\ F_{41} & F_{42} & F_{43} & 0 \end{vmatrix}$$

are of much importance. Let us write,

$$(70)\ fF = \begin{vmatrix} S_{11} - L & S_{12} & S_{13} & S_{14} \\ S_{21} & S_{22} - L & S_{23} & S_{24} \\ S_{31} & S_{32} & S_{33} - L & S_{34} \\ S_{41} & S_{42} & S_{43} & S_{44} - L \end{vmatrix}$$

Then (71) $S_{11} + S_{22} + S_{33} + S_{44} = 0.$

Let L now denote the symmetrical combination of the indices 1, 2, 3, 4, given by

$$(72)\ L = \tfrac{1}{2}(f_{23} F_{23} + f_{31}F_{31} + f_{12} + F_{12} + f_{14} F_{14} + f_{24} F_{24} + f_{34} F_{34})$$

Then we shall have

$$(73)\ S_{11} = \tfrac{1}{2}(f_{23} F_{23} + f_{34} F_{34} + f_{42} F_{42} - f_{12} F_{12} - f_{13} F_{13} f_{14} F_{14})$$

$S_{12} = f_{13} F_{32} + f_{14} F_{42}$ etc....

In order to express in a real form, we write

(74) $S = | S_{11}\ S_{12}\ S_{13}\ S_{14} |$

$| S_{21}\ S_{22}\ S_{23}\ S_{24} |$

$| S_{31}\ S_{32}\ S_{33}\ S_{34} |$

$| S_{41}\ S_{42}\ S_{43}\ S_{44} |$

$= | X_x\ Y_x\ Z_x\ -iT_x |$

$| X_y\ Y_y\ Z_y\ -iT_y |$

$| X_z\ Y_z\ Z_z\ -iT_z |$

$| -iX_t\ -iY_t\ -iZ_t\ T_t |$

Now $X_x = \frac{1}{2}[m_xM_x - m_yM_y - m_zM_z + e_xE_x - e_yE_y - e_zE_z]$

so

(75) $X_y = m_xM_y + e_yE_x$, $Y_x = m_yM_x + e_xE_y$ etc.

$X_t = e_yM_z - e_zM_y$, $T_x = m_xE_y - m_yE_z$, etc.

$T_t = \frac{1}{2}[m_xM_x + m_yM_y + m_zM_z + e_xE_x + e_yE_y + e_zE_z]$

$L_t = \frac{1}{2}[m_xM_x + m_yM_y + m_zM_z - e_xE_x - e_yE_y - e_zE_z]$

These quantities[27] are all real. In the theory for bodies at rest, the combinations (X_x, X_y, X_z, Y_z, Y_y, Y_z, Z_x, Z_y, Z_z) are known as "Maxwell's Stresses," T_x, T_y, T_z are known as the Poynting's Vector, T_t as the electromagnetic energy-density, and L as the Langrangian function.

On the other hand, by multiplying the alternating matrices of f^* and F^*, we obtain

(77) $F^*f^* = | -S_{11} - L, -S_{12}, -S_{13}. -S_{14} |$

$| -S_{21}, -S_{22} - L, -S_{23}, -S_{24} |$

$| -S_{31}\ -S_{32}, -S_{33} - L, -S_{34} |$

$| -S_{41}\ -S_{42}\ -S_{43}\ -S_{44} - L |$

and hence, we can put

(78) $fF = S - L, F^*f^* = -S - L,$

where by L, we mean L-times the unit matrix, *i.e.* the matrix with elements

$| Le_{hk} |, (e_{hh} = 1, e_{hk} = 0, h \neq k\ h, k = 1, 2, 3, 4).$

Since here SL = LS, we deduce that,

$F^*f^*fF = (-S - L)(S - L) = -SS + L^2,$

and find, since $f^*f = \mathrm{Det}^{1/2}f$, $F^*F = \mathrm{Det}^{1/2}F$, we arrive at the interesting conclusion

(79) $SS = L^2 - \mathrm{Det}^{1/2}f\,\mathrm{Det}^{1/2}F$

i.e. the product of the matrix S into itself can be expressed as the multiple of a unit matrix—a matrix in which all the elements except those in the principal diagonal are zero, the elements in the principal diagonal are all equal and have the value given on the right-hand side of (79). Therefore the general relations

(80) $S_{h1}\,S_{1k} + S_{h2}\,S_{2k} + S_{h3}\,S_{3k} + S_{h4}\,S_{4k} = 0,$

h, *k* being unequal indices in the series 1, 2, 3, 4, and

(81) $S_{h1}\,S_{1h} + S_{h2}\,S_{2h} + S_{h3}\,S_{3h} + S\{h4\}\,S_{4h} = L^2 -$

$\mathrm{Det}^{1/2}f\,\mathrm{Det}^{1/2}F,$

for $h = 1, 2, 3, 4$.

Now if instead of F, and *f* in the combinations (72) and (73), we introduce the electrical rest-force Φ, the magnetic rest-force ψ, and the rest-ray Ω [(55), (56) and (57)], we can pass over to the expressions,—

(82) $L = -\frac{1}{2}\,\varepsilon\,\Phi\,[=\Phi] + \frac{1}{2}\,\mu\,\psi\,[=\psi],$

(83) $S_{hk} = -\frac{1}{2}\,\varepsilon\,\Phi\,[=\Phi]\,e_{hk} - \frac{1}{2}\,\mu\,\psi\,[=\psi]\,e_{hk}$

$+ \varepsilon (\Phi_h \Phi_k - \Phi ([=\Phi]) \omega_h \Omega_k$

$+ \mu (\psi_h \psi_k - \Psi [=\psi] \Omega\{h\} \omega_k) - \omega_h \omega_k - \varepsilon\mu \omega_h \Omega_k$

$(h_1 k = 1, 2, 3, 4)$.

Here we have

$$\Phi [=\Phi] = \Phi_1^2 + \Phi_2^2 + \Phi_3^2 + \Phi_4^2, \psi[=\psi] = \psi_1^2 + \psi_2^2 + \psi_3^2 + \psi_4^2$$

$e_{hh} = 1, e_{hk} = 0 \ (h \neq k)$.

The right side of (82) as well as L is an invariant in a Lorentz transformation, and the 4 × 4 element on the right side of (83) as well as $S_{k h}$ represent a space time vector of the second kind. Remembering this fact, it suffices, for establishing the theorems (82) and (83) generally, to prove it for the special case $\omega_1 = 0, \omega_2 = 0, \omega_3 = 0, \omega_4 = i$. But for this case $\omega = 0$, we immediately arrive at the equations (82) and (83) by means (45), (51), (60) on the one hand, and $e = \varepsilon E$, $M = \mu m$ on the other hand.

The expression on the right-hand side of (81), which equals

$[\frac{1}{2} (m M - eE)^2] + (em) (EM)$,

is >= 0, because ($em = \varepsilon \Phi [=\psi]$, $(EM) = \mu \Phi [=\psi]$; now referring back to 79), we can denote the positive square root of this expression as $\mathrm{Det}^{1/4} S$.

Since $\square = -f$, and $\dot{F} = -F$, we obtain for $\dot{S}$, the transposed matrix of S, the following relations from (78),

(84) $Ff = \dot{S} - L, f^* F^* = -\dot{S} - L$,

Then is

$\dot{S} - S = | S_{hk} - S_{tk} |$

an alternating matrix, and denotes a space-time vector of the second kind. From the expressions (83), we obtain,

(85) $S - \dot{S} = - (\varepsilon\mu - 1) [\omega, \Omega]$,

from which we deduce that [see (57), (58)].

(86) $\omega (S - \dot{S})^* = 0$,

(87) ω (S - $\dot{S}$) = ($\varepsilon\mu$ - 1) Ω

When the matter is at rest at a space-time point, ω = 0, then the equation 86) denotes the existence of the following equations

$Z_y = Y_z, X_z = Z_x, Y_x = X_y,$

and from 83),

$T_x = \Omega_1, T_y = \Omega_2, T_z = \Omega_3$

$X_t = \varepsilon\mu\Omega_1, Y_t = \varepsilon\mu\Omega_2, Z_t = \varepsilon\mu\Omega_3$

Now by means of a rotation of the space co-ordinate system round the null-point, we can make,

$Z_y = Y_z = 0, X_z = Z_x = 0, X_x = X_y = 0,$

According to 71), we have

(88) $X_x + Y_y + Z_z + T_t = 0,$

and according to 83), $T_t > 0$. In special cases, where ω vanishes it follows from 81) that

$X_x^2 = Y_y^2 = Z_z^2 = T_t^2, = (\mathrm{Det}^{1/4}\ S)^2,$

and if T, and one of the three magnitudes X_x, Y_y, Z_z are = $\pm\mathrm{Det}^{1/4}$ S, the two others = -$\mathrm{Det}^{1/4}$ S. If Ω does not vanish let $\Omega \neq 0$, then we have in particular from 80)

$T_z X_t = 0, T_z Y_t = 0, Z_z T_z + T_z T_t = 0,$

and if $\Omega_1 = 0, \Omega_2 = 0, Z_z = -T_t$ It follows from (81), (see also 83) that

$X_x = -Y_y = \pm\mathrm{Det}^{1/4}\ S,$

and $-Z_z = T_t = \sqrt{(\mathrm{Det}^{1/2}\ S + \varepsilon\mu\Omega_3^2)} > \mathrm{Det}^{1/4}S.$

The space-time vector of the first kind

(89) K = lor S,

is of very great importance for which we now want to demonstrate a very important transformation

According to 78), $S = L + fF$, and it follows that

$$\text{lor } S = \text{lor } L + \text{lor } fF.$$

The symbol 'lor' denotes a differential process which in lor fF, operates on the one hand upon the components of f, on the other hand also upon the components of F. Accordingly lor fF can be expressed as the sum of two parts. The first part is the product of the matrices (lor f) F, lor f being regarded as a 1×4 series matrix. The second part is that part of lor fF, in which the diffentiations operate upon the components of F alone. From 78) we obtain

$$fF = -F^*f^* - 2L;$$

hence the second part of lor fF = -(lor F^*)f^* + the part of -2 lor L, in which the differentiations operate upon the components of F alone. We thus obtain

$$\text{lor } S = (\text{lor } f)F - (\text{lor } F^*)f^* + N,$$

where N is the vector with the components

$$N_h = \tfrac{1}{2}(\partial f_{23}/\partial x_h F_{23} + \partial f_{31}/\partial x_h F_{31} + \partial f_{12}/\partial x_h F_{12} + \partial f_{14}/\partial x_h F_{14}$$

$$+ \partial f_{24}/\partial x_h F_{24} + \partial f_{34}/\partial x_h F_{34}$$

$$- \partial F_{23}/\partial x_h f_{23} - \partial F_{31}/\partial x_h f_{31} - \partial F_{12}/\partial x_h f_{12} - \partial F_{14}/\partial x_h f_{14}$$

$$- \partial F_{24}/\partial x_h f_{24} - \partial F_{34}/\partial x_h f_{34}),$$

$$(h = 1, 2, 3, 4)$$

By using the fundamental relations A) and B), 90) is transformed into the fundamental relation

$$(91) \quad \text{lor } S = -sF + N.$$

In the limitting case $\varepsilon = 1$, $\mu = 1$, $f = F$, N vanishes identically.

Now upon the basis of the equations (55) and (56), and referring back to the expression (82) for L, and from 57) we obtain the following expressions as components of N,—

$$(92)\ N_h = -\tfrac{1}{2}\,\Phi[=\Phi]\partial\varepsilon/\partial x_h - \tfrac{1}{2}\,\psi[=\psi]\partial\mu/\partial x_h$$

$$+ (\varepsilon\mu - 1)(\Omega_1\,\partial\omega_1/\partial x_h + \Omega_2\,\partial\omega_2/\partial x_h + \Omega_3\,\partial\omega_3/\partial x_h + \Omega_4\,\partial\omega_4/\partial x_h)$$

for $h = 1, 2, 3, 4$.

Now if we make use of (59), and denote the space-vector which has Ω_1, Ω_2, Ω_3 as the x, y, z components by the symbol W, then the third component of 92) can be expressed in the form

$$(93)\ (\varepsilon\mu - 1)/\sqrt{(1 - u^2)}\ (W\,\partial u/\partial x_h),$$

The round bracket denoting the scalar product of the vectors within it.

§ 14. The Ponderomotive Force.[28]

Let us now write out the relation K = lor S = -sF + N in a more practical form; we have the four equations

(94) $K_1 = \partial X_x/\partial x + \partial X_y/\partial y + \partial X_y/\partial z - \partial X_t/\partial t = \varrho E_x + s_y M_z - s_z M_x$

$- \frac{1}{2} \Phi[=\Phi]\, \partial\varepsilon/\partial x - \frac{1}{2} \psi[=\psi]\partial\mu/\partial x + (\varepsilon\mu - 1)/\sqrt{(1 - u^2)}\, (W\partial u/\partial x)$,

(95) $K_2 = \partial Y_x/\partial x + \partial Y_y/\partial y + \partial Y_z/\partial z - \partial Y_t/\partial t = \varrho E_y + s_z M_x - s_x M_y$

$- \frac{1}{2} \Phi[=\Phi]\partial\varepsilon/\partial y - \frac{1}{2} \psi[=\psi]\partial\mu/\partial y + (\varepsilon\mu - 1)/\sqrt{(1 - u^2)}\, (W\partial u/\partial y)$,

(96) $K_3 = \partial Z_x/\partial x + \partial Z_y/\partial y + \partial Z_z/\partial z - \partial Z_t/\partial t = \varrho E_2 + s_x M_y - s_y M_4$

$- \frac{1}{2} \Phi[=\Phi]\, \partial\varepsilon/\partial z - \frac{1}{2} \psi[=\psi]\, \partial\mu/\partial z + (\varepsilon\mu - 1)/\sqrt{(1 - u^2)}\, (W\partial u/\partial z)$,

(97) $(1/i)K_4 = \partial T_y/\partial x - \partial T_y/\partial y - \partial T_z/\partial z - \partial T_t/\partial t = s_x E_x + s_y E_y + s_z E_z$

$- \frac{1}{2} \Phi[=\Phi]\partial\varepsilon/\partial t - \frac{1}{2} \psi[=\psi]\partial\mu/\partial t + (\varepsilon\mu - 1)/\sqrt{(1 - u^2)}\, (W\partial u/\partial t)$.

It is my opinion that when we calculate the ponderomotive force which acts upon a unit volume at the space-time point x, y, z, t, it has got, x, y, z components as the first three components of the space-time vector

K + (ωK)ω,

This vector is perpendicular to ω; the law of Energy finds its expression in the fourth relation.

The establishment of this opinion is reserved for a separate tract.

In the limiting case ε = 1, μ = 1, σ = 0, the vector N = 0, S = ϱω, ωK = 0, and we obtain the ordinary equations in the theory of electrons.

APPENDIX
Mechanics and the Relativity-Postulate.

It would be very unsatisfactory, if the new way of looking at the time-concept, which permits a Lorentz transformation, were to be confined to a single part of Physics.

Now many authors say that classical mechanics stand in opposition to the relativity postulate, which is taken to be the basis of the new Electro-dynamics.

In order to decide this let us fix our attention upon a special Lorentz transformation represented by (10), (11), (12), with a vector v in any direction and of any magnitude $q < 1$ but different from zero. For a moment we shall not suppose any special relation to hold between the unit of length and the unit of time, so that instead of t, t', q, we shall write ct, ct', and q/c, where c represents a certain positive constant, and q is $< c$. The above mentioned equations are transformed into

$$r'_{\Box} = r_{\Box},$$

$$r'_v = c(r_v - qt)/\sqrt{(c^2 - q^2)},$$

$$t' = (qr_v + c^2t)/c\sqrt{(c^2 - q^2)}$$

They denote, as we remember, that r is the space-vector (x, y, z), r' is the space-vector $(x' y' z')$

If in these equations, keeping v constant we approach the limit $c = \infty$, then we obtain from these

$$r'_{\Box} = r_{\Box},$$

$$r'_v = r_v - qt,$$

$$t' = t.$$

The new equations would now denote the transformation of a spatial co-ordinate system (x, y, z) to another spatial co-ordinate system $(x' y' z')$ with parallel axes, the null point of the second system moving with constant velocity in a straight line, while the time parameter remains unchanged. We can, therefore, say that classical mechanics postulates a covariance of Physical laws for the group of homogeneous linear transformations of the expression

$-x^2 - y^2 - z^2 + c^2$ (1)

when $c = \infty$.

Now it is rather confusing to find that in one branch of Physics, we shall find a covariance of the laws for the transformation of expression (1) with a finite value of c, in another part for $c = \infty$.

It is evident that according to Newtonian Mechanics, this covariance holds for $c = \infty$ and not for c = velocity of light.

May we not then regard those traditional covariances for $c = \infty$ only as an approximation consistent with experience, the actual covariance of natural laws holding for a certain finite value of c.

I may here point out that by if instead of the Newtonian Relativity-Postulate with $c = \infty$, we assume a relativity-postulate with a finite c, then the axiomatic construction of Mechanics appears to gain considerably in perfection.

The ratio of the time unit to the length unit is chosen in a manner so as to make the velocity of light equivalent to unity.

While now I want to introduce geometrical figures in the manifold of the variables (x, y, z, t), it may be convenient to leave (y, z) out of account, and to treat x and t as any possible pair of co-ordinates in a plane, referred to oblique axes.

A space time null point 0 $(x, y, z, t = 0, 0, 0, 0)$ will be kept fixed in a Lorentz transformation.

The figure $-x^2 - y^2 - z^2 + t^2 = 1, t > 0$ ... (2)

which represents a hyper boloidal shell, contains the space-time points A $(x, y, z, t = 0, 0, 0, 1)$, and all points A' which after a Lorentz-transformation enter into the newly introduced system of reference as $(x', y', z', t' = 0, 0, 0, 1)$.

The direction of a radius vector 0A' drawn from 0 to the point A' of (2), and the directions of the tangents to (2) at A' are to be called normal to each other.

Let us now follow a definite position of matter in its course through all time t. The totality of the space-time points (x, y, z, t) which correspond to the positions at different times t, shall be called a space-time line.

The task of determining the motion of matter is comprised in the following problem:—It is required to establish for every space-time point the direction of the space-time line passing through it.

To transform a space-time point P (x, y, z, t) to rest is equivalent to introducing, by means of a Lorentz transformation, a new system of reference (x', y', z', t'), in which the t' axis has the direction 0A', 0A' indicating the direction of the space-time line passing through P. The space $t' =$ const, which is to be laid through P, is the one which is perpendicular to the space-time line through P.

To the increment dt of the time of P corresponds the increment

$$d\tau = \sqrt{(dt^2 - dx^2 - dy^2) - dz^2} = dt\sqrt{(1 - u^2)}$$

of the newly introduced time parameter t'. The value of the integral

$$\int d\tau = \int \sqrt{(-(dx_1^2 + dx_2^2 + dx_3^2 + dx_4^2))}$$

when calculated upon the space-time line from a fixed initial point P_0 to the variable point P, (both being on the space-time line), is known as the 'Proper-time' of the position of matter we are concerned with at the space-time point P. (It is a generalization of the idea of Positional-time which was introduced by Lorentz for uniform motion.)

If we take a body R_0 which has got extension in space at time t_0, then the region comprising all the space-time line passing through R_0 and t_0 shall be called a space-time filament.

If we have an analytical expression $\theta(x\ y, z, t)$ so that $\theta(x, y\ z\ t) = 0$ is intersected by every space time line of the filament at one point,—whereby

$$-(\partial\Theta/\partial x)^2, -(\partial\Theta/\partial y)^2, -(\partial\Theta/\partial z)^2,$$

$$-(\partial\Theta/\partial t)^2 > 0, \partial\Theta/\partial t > 0.$$

then the totality of the intersecting points will be called a cross section of the filament.

At any point P of such across-section, we can introduce by means of a Lorentz transformation a system of reference $(x', y, z'\ t)$, so that according to this

$$\partial\Theta/\partial x' = 0, \partial\Theta/\partial y' = 0, \partial\Theta/\partial z' = 0, \partial\Theta/\partial t' > 0.$$

The direction of the uniquely determined t'—axis in question here is known as the upper normal of the cross-section at the point P and the value of $dJ = \iiint dx'\ dy'\ dz'$ for the surrounding points of P on the cross-section is known as the elementary contents (Inhalts-element) of the cross-section. In this sense R_0 is to be regarded as the cross-section normal to the t axis of the filament at the point $t = t_0$, and the volume of the body R_0 is to be regarded as the contents of the cross-section.

If we allow R_0 to converge to a point, we come to the conception of an infinitely thin space-time filament. In such a case, a space-time line will be thought of as a principal line and by the term 'Proper-time' of the filament will be understood the 'Proper-time' which is laid along this principal line; under the term normal cross-section of the filament, we shall understand the cross-section upon the space which is normal to the principal line through P.

We shall now formulate the principle of conservation of mass.

To every space R at a time t, belongs a positive quantity—the mass at R at the time t. If R converges to a point (x, y, z, t), then the quotient of this mass, and the volume of R approaches a limit $\mu(x, y, z, t)$, which is known as the mass-density at the space-time point (x, y, z, t).

The principle of conservation of mass says—that for an infinitely thin space-time filament, the product μdJ, where μ = mass-density at the point (x, y, z, t) of the filament (*i.e.*, the principal line of the filament), dJ = contents of the cross-section normal to the t axis, and passing through (x, y, z, t), is constant along the whole filament.

Now the contents dJ_n of the normal cross-section of the filament which is laid through (x, y, z, t) is

$$(4)\ dJ_n = (1/\sqrt{(1 - u^2)})dJ = -i\omega_4\ dJ = (dt/d\tau)dJ.$$

and the function

$$\nu = \mu/-i\omega_4 = \mu\sqrt{(1 - u^2)}) = \mu(\partial\tau/\partial t.\ (5)$$

may be defined as the rest-mass density at the position $(x\ y\ z\ t)$. Then the principle of conservation of mass can be formulated in this manner:—

For an infinitely thin space-time filament, the product of the rest-mass density and the contents of the normal cross-section is constant along the whole filament.

In any space-time filament, let us consider two cross-sections $Q°$ and Q', which have only the points on the boundary common to each other; let the space-time lines inside the filament have a larger value of t on Q' than on Q°. The finite range enclosed between Q° and Q' shall be called a space-time *sichel*,[29] Q' is the lower boundary, and Q' is the upper boundary of the *sichel.*

If we decompose a filament into elementary space-time filaments, then to an entrance-point of an elementary filament through the lower boundary of the *sichel*, there corresponds an exit point of the same by the upper boundary, whereby for both, the product νdJ_n taken in the sense of (4) and (5), has got the same value. Therefore the difference of the two integrals $\int \nu dJ_n$ (the first being extended over the upper, the second upon the lower boundary) vanishes. According to a well-known theorem of Integral Calculus the difference is equivalent to

$$\iiiint \text{lor } \nu[=\omega]\, dx\, dy\, dz\, dt,$$

the integration being extended over the whole range of the *sichel*, and (comp. (67), § 12)

$$\text{lor } \nu[=\omega] = (\partial\nu\omega_1/\partial x_1) + (\partial\nu\omega_2/\partial x_2) + (\partial\nu\omega_3/\partial x_3) + (\partial\nu\omega_4/\partial x_4).$$

If the *sichel* reduces to a point, then the differential equation

$$\text{lor } \nu[=\omega] = 0, \quad (6)$$

which is the condition of continuity

$$(\partial\mu u_x/\partial x) + (\partial\mu u_y/\partial y) + (\partial\mu u_z/\partial z) + (\partial\mu/\partial t) = 0.$$

Further let us form the integral

$$N = \int\iiint \nu\, dx\, dy\, dz\, dt \quad (7)$$

extending over the whole range of the space-time *sichel.* We shall decompose the *sichel* into elementary space-time filaments, and every one of these filaments in small elements $d\tau$ of its proper-time, which are however large compared to the linear dimensions of the normal cross-section; let us assume that the mass of such a filament $\nu dJ_n = dm$ and write τ^0, τ^1 for the 'Proper-time' of the upper and lower boundary of the *sichel.*

Then the integral (7) can be denoted by

$$\iint \nu dJ_n\, d\tau = \int (\tau' - \tau^0)\, dm.$$

taken over all the elements of the sichel.

Now let us conceive of the space-time lines inside a space-time *sichel* as material curves composed of material points, and let us suppose that they are subjected to a continual change of length inside the sichel in the following manner. The entire curves are to be varied in any possible manner inside the *sichel*, while the end points on the lower and upper boundaries remain fixed, and the individual substantial points upon it are displaced in such a manner that they always move forward normal to the curves. The whole process may be analytically represented by means of a parameter λ, and to the value $\lambda = 0$, shall correspond the actual curves inside the *sichel*. Such a process may be called a virtual displacement in the sichel.

Let the point (x, y, z, t) in the sichel $\lambda = 0$ have the values $x + \delta x, y + \delta y, z + \delta z, t + \delta t$, when the parameter has the value λ; these magnitudes are then functions of (x, y, z, t, λ). Let us now conceive of an infinitely thin space-time filament at the point $(x\ y\ z\ t)$ with the normal section of contents dJ_n and if $dJ_n + \delta dJ_n$ be the contents of the normal section at the corresponding position of the varied filament, then according to the principle of conservation of mass—($\nu + d\nu$ being the rest-mass-density at the varied position),

(8) $(\nu + \delta\nu)\ (dJ_n + \delta dJ_n) = \nu dJ_n = dm.$

In consequence of this condition, the integral (7) taken over the whole range of the *sichel*, varies on account of the displacement as a definite function $N + \delta N$ of λ, and we may call this function $N + \delta N$ as the *mass action* of the virtual displacement.

If we now introduce the method of writing with indices, we shall have

(9) $d(x_h + \delta x_h) = dx_h + \sum_k \partial\delta x_h/\partial x_k + \partial\delta x_h/\partial\lambda\ d\lambda$

$k = 1, 2, 3, 4$

$h = 1, 2, 3, 4$

Now on the basis of the remarks already made, it is clear that the value of $N + \delta N$, when the value of the parameter is λ, will be:—

(10) $N + \delta N = \iiiint ((\nu d(\tau + \delta\tau))/d\tau) dx\ dy\ dz\ dt,$

the integration extending over the whole sichel $d(\tau + \delta\tau)$ where $d(\tau + \delta\tau)$ denotes the magnitude, which is deduced from

$\sqrt{(-(dx_1 + d\delta x_1)^2 - (dx_2 + d\delta x_2)^2 - (dx_3 + d\delta x_3)^2 - (dx_4 + d\delta x_4)^2)}$

by means of (9) and

$dx_1 = \omega_1\, d\tau,\ dx_2 = \omega_2\, d\tau,$

$dx_3 = \omega_3\, d\tau,\ dx_4 = \omega_4\, d\tau,\ d\lambda = 0$

therefore:—

(11) $(d(\tau + \delta\tau))/d\tau = \sqrt{(-\sum(\omega_h + \sum(\partial\delta x_h/\partial x_k)\omega_k)^2)}$

$k = 1, 2, 3, 4.$

$h = 1, 2, 3, 4.$

We shall now subject the value of the differential quotient

(12) $((d(N + \delta N))/d\lambda)\ (\lambda = 0)$

to a transformation. Since each δx_h as a function of (x, y, z, t) vanishes for the zero-value of the parameter λ, so in general $d\delta x_k/(\partial x_h = 0$, for $\lambda = 0$.

Let us now put $(\partial\delta x_h/\partial\lambda) = \xi_h\ (h = 1, 2, 3, 4)$ (13)

$\lambda = 0$

then on the basis of (10) and (11), we have the expression (12):—

$= -\iiiint \sum \omega_h((\partial\xi_h/\partial x_1)\omega_1 + (\partial\xi_h/\partial x_2)\omega_2 + (\partial\xi_h/\partial x_3)\omega_3 + (\partial\xi_h/\partial x_4)\omega_4)$

$dx\, dy\, dz\, dt$

for the system $(x_1\ x_2\ x_3\ x_4)$ on the boundary of the *sichel*, $(\delta x_1\ \delta x_2\ \delta x_3\ \delta x_4)$ shall vanish for every value of λ and therefore $\xi_1, \xi_2, \xi_3, \xi_4$ are nil. Then by partial integration, the integral is transformed into the form

$\iiiint \sum \xi_h(\partial\nu\omega_h\omega_1/\partial x_1 + \partial\nu\omega_h\omega_2/\partial x_2 + \partial\nu\omega_h\omega_3/\partial x_3 + \partial\nu\omega_h\omega_4/\partial x_4)$

$dx\, dy\, dz\, dt$

the expression within the bracket may be written as

$= \omega_h \sum \partial\nu\omega_k/\partial x_k + \nu\sum\omega_k\partial\omega_h/\partial x_k.$

The first sum vanishes in consequence of the continuity equation (b). The second may be written as

$(\partial\omega_h/\partial x_1)(dx_1/d\tau) + (\partial\omega_h/\partial x_2)(dx_2/d\tau) + (\partial\omega_h/\partial x_3)(dx_3/d\tau) + (\partial\omega_h/\partial x_4)(dx_4/d\tau)$

$= d\omega_h/d\tau = (d/d\tau)(dx_h/d\tau)$

whereby $(d/d\tau)$ is meant the differential quotient in the direction of the space-time line at any position. For the differential quotient (12), we obtain the final expression

(14) $\iiiint \nu((\partial\omega_1/\partial\tau)\xi_1 + (\partial\omega_2/\partial\tau)\xi_2 + (\partial\omega_3/\partial\tau)\xi_3 + (\partial\omega_4/\partial\tau)\xi_4)$

$dx\ dy\ dz\ dt.$

For a virtual displacement in the *sichel* we have postulated the condition that the points supposed to be substantial shall advance normally to the curves giving their actual motion, which is $\lambda = 0$; this condition denotes that the ξ_h is to satisfy the condition

$w_1\xi_1 + w_2\xi_2 + w_3\xi_3 + w_4\xi_4 = 0.$ (15)

Let us now turn our attention to the Maxwellian tensions in the electrodynamics of stationary bodies, and let us consider the results in § 12 and 13; then we find that Hamilton's Principle can be reconciled to the relativity postulate for continuously extended elastic media.

At every space-time point (as in § 13), let a space time matrix of the 2nd kind be known

(16) S =

$| S_{11}\ S_{12}\ S_{13}\ S_{14} | = | X_x\ Y_x\ Z_x\ -iT_x |$

$| S_{21}\ S_{22}\ S_{23}\ S_{24} | = | X_y\ Y_y\ Z_y\ -iT_y |$

$| S_{31}\ S_{32}\ S_{33}\ S_{34} | = | X_z\ Y_z\ Z_z\ -iT_z |$

$| S_{41}\ S_{42}\ S_{43}\ S_{44} | = | -iX_t\ -iY_t\ -iZ_t\ T_t |$

where X_n, Y_x .....X_z, T_t are real magnitudes.

For a virtual displacement in a space-time sichel (with the previously applied designation) the value of the integral

(17) $W + \delta W = \iiiint (\sum S_{hk}\ (\partial(x_k + \delta x_k))/\partial x_h\ dx\ dy\ dz\ dt$

extended over the whole range of the *sichel*, may be called the tensional work of the virtual displacement.

The sum which comes forth here, written in real magnitudes, is

$$X_x + Y_y + Z_z + T_t + X_x (\partial\delta x)/\partial x + X_y (\partial\delta x)/\partial y + \ldots Z_z (\partial\delta z)/\partial z$$

$$- X_t (\partial\delta x/\partial t - \ldots + T_x (\partial\delta t)/\partial x + \ldots T_t (\partial\delta t)/\partial t$$

we can now postulate the following *minimum principle in mechanics.*

If any space-time Sichel be bounded, then for each virtual displacement in the Sichel, the sum of the mass-works, and tension works shall always be an extremum for that process of the space-time line in the Sichel which actually occurs.

The meaning is, that for each virtual displacement,

$$([d(\cdot\delta N + \delta W)]/d\lambda)_{\lambda = 0} = 0 \ (18)$$

By applying the methods of the Calculus of Variations, the following four differential equations at once follow from this minimal principle by means of the transformation (14), and the condition (15).

$$(19)\ \nu\ \partial w_h/\partial\tau = K_h + \chi w_h\ (h = 1, 2, 3, 4)$$

$$\text{whence } K_h = \partial S_{1\,h}/\partial x_1 + \partial S_{2\,h}/\partial x_2 + \partial S_{3\,h}/\partial x_3 + \partial S_{4\,h}/\partial x_4,\ (20)$$

are components of the space-time vector 1st kind K = lor S, and X is a factor, which is to be determined from the relation $w\Box = -1$. By multiplying (19) by w_h, and summing the four, we obtain X = K□, and therefore clearly K + (K□)w will be a space-time vector of the 1st kind which is normal to w. Let us write out the components of this vector as

$$X, Y, Z, \cdot iT$$

Then we arrive at the following equation for the motion of matter,

$$(21)\ \nu\ d/d\tau\ (dx/d\tau) = X,\ \nu\ d/d\tau\ (dy/d\tau) = Y,\ \nu\ d/d\tau\ (dz/d\tau) = Z,$$

$\nu\ d/d\tau\ (dx/d\tau) = T$, and we have also

$$(dx/d\tau)^2 + (dy/d\tau)^2 + (dz/d\tau)^2 > (dt/d\tau)^2 = -1,$$

$$\text{and } X\ dx/d\tau + Y\ dy/d\tau + Z\ dz/d\tau = T\ dt/d\tau.$$

On the basis of this condition, the fourth of equations (21) is to be regarded as a direct consequence of the first three.

From (21), we can deduce the law for the motion of a material point, *i.e.*, the law for the career of an infinitely thin space-time filament.

Let x, y, z, t, denote a point on a principal line chosen in any manner within the filament. We shall form the equations (21) for the points of the normal cross section of the filament through x, y, z, t, and integrate them, multiplying by the elementary contents of the cross section over the whole space of the normal section. If the integrals of the right side be R_x R_y R_z R_t and if m be the constant mass of the filament, we obtain

(22) $m\, d/d\tau\, dx/d\tau = R_x$,

$m\, d/d\tau\, dy/d\tau = R_y$,

$m\, d/d\tau\, dz/d\tau = R_z$,

$m\, d/d\tau\, dt/d\tau = R_t$

R is now a space-time vector of the 1st kind with the components (R_x R_y R_z R_t) which is normal to the space-time vector of the 1st kind w,—the velocity of the material point with the components

$dx/d\tau,\ dy/d\tau,\ dz/d\tau,\ i\, dt/d\tau$.

We may call this vector R *the moving force of the material point.*

If instead of integrating over the normal section, we integrate the equations over that cross section of the filament which is normal to the t axis, and passes through (x, y, z, t), then [See (4)] the equations (22) are obtained, but

are now multiplied by $d\tau/dt$; in particular, the last equation comes out in the form,

$m\, d/dt\, (dt/d\tau) = w_x R_x\, d\tau/dt + w_y R_y\, d\tau/dt + w_z R_z\, d\tau/dt$.

The right side is to be looked upon *as the amount of work done per unit of time* at the material point. In this equation, we obtain the energy-law for the motion of the material point and the expression

$m\, (dt/d\tau - 1) = m\, [1/\sqrt{(1 - w^2)} - 1]$

$= m\, (½\ |w_1{}^2 + 3/8\ |w_1{}^4 +)$

may be called the kinetic energy of the material point.

Since dt is always greater than $d\tau$ we may call the quotient $(dt - d\tau)/d\tau$ as the "Gain" (vorgehen) of the time over the proper-time of the material point and the law can then be thus expressed;—The kinetic energy of a material point is the product of its mass into the gain of the time over its proper-time.

The set of four equations (22) again shows the symmetry in (x, y, z, t), which is demanded by the relativity postulate; to the fourth equation however, a higher physical significance is to be attached, as we have already seen in the analogous case in electrodynamics. On the ground of this demand for symmetry, the triplet consisting of the first three equations are to be constructed after the model of the fourth; remembering this circumstance, we are justified in saying,—

"If the relativity-postulate be placed at the head of mechanics, then the whole set of laws of motion follows from the law of energy."

I cannot refrain from showing that no contradiction to the assumption on the relativity-postulate can be expected from the phenomena of gravitation.

If $B^*(x^*, y^*, z^*, t^*)$ be a solid (fester) space-time point, then the region of all those space-time points $B\ (x, y, z, t)$, for which

$$(23)\quad (x - x^*)^2 + (y - y^*)^2 + (z - z^*)^2 = (t - t^*)^2$$

$$t - t^* >= 0$$

may be called a "Ray-figure" (Strahl-gebilde) of the space time point B*.

A space-time line taken in any manner can be cut by this figure only at one particular point; this easily follows from the convexity of the figure on the one hand, and on the other hand from the fact that all directions of the space-time lines are only directions from B* towards to the concave side of the figure. Then B* may be called the light-point of B.

If in (23), the point $(x\ y\ z\ t)$ be supposed to be fixed, the point $(x^*\ y^*\ z^*\ t^*)$ be supposed to be variable, then the relation (23) would represent the locus of all the space-time points B*, which are light-points of B.

Let us conceive that a material point F of mass m may, owing to the presence of another material point F*, experience a moving force according to the following law. Let us picture to ourselves the space-time filaments of F and F* along with the principal lines of the filaments. Let BC be an infinitely small element of the principal line of F; further let B* be the light point of B, C* be the light point of C on the principal line of F*; so that OA' is the radius vector of the hyperboloidal fundamental figure (23) parallel to B*C*, finally D* is the point of intersection of line B*C* with the space normal to itself and passing through B. The moving force of the mass-point F in the space-time point B is now the space-time vector of the first kind which is normal to BC, and which is composed of the vectors

(24) $mm^*(OA'/B^*D^*)^3$ BD* in the direction of BD*, and another vector of suitable value in direction of B*C*.

Now by (OA'/B*D*) is to be understood the ratio of the two vectors in question. It is clear that this proposition at once shows the covariant character with respect to a Lorentz-group.

Let us now ask how the space-time filament of F behaves when the material point F* has a uniform translatory motion, *i.e.*, the principal line of the filament of F* is a line. Let us take the space time null-point in this, and by means of a Lorentz-transformation, we can take this axis as the t-axis. Let x, y, z, t, denote the point B, let τ^* denote the proper time of B*, reckoned from O. Our proposition leads to the equations

(25) $d^2x/d\tau^2 = - m^*x/(t - \tau^*)^2, d^2y/d\tau^2 = - m^*y/(t - \tau^*)^3$

$d^2z/d\tau^2 = -m^*z/(t - \tau^*)^3,$

(26) $d^2t/d\tau^2 = -m^*/(t - \tau^*)^2 \, d(t - \tau^*)/dt$

where (27) $x^2 + y^2 + z^2 = (t - \tau^*)^2$

and (28) $(dx/d\tau)^2 + (dy/d\tau)^2 + (dz/d\tau)^2 = (dt/d\tau)^2 - 1.$

In consideration of (27), the three equations (25) are of the same form as the equations for the motion of a material point subjected to attraction from a fixed centre according to the Newtonian Law, only that instead of the time t, the proper time τ of the material point occurs. The fourth equation (26) gives then the connection between proper time and the time for the material point.

Now for different values of τ', the orbit of the space-point $(x\ y\ z)$ is an ellipse with the semi-major axis a and the eccentricity e. Let E denote the eccentric anomaly, T the increment of the proper time for a complete description of the orbit, finally $n\mathrm{T} = 2\pi$, so that from a properly chosen initial point τ, we have the Kepler-equation

(29) $n\tau = \mathrm{E} - e \sin \mathrm{E}.$

If we now change the unit of time, and denote the velocity of light by c, then from (28), we obtain

(30) $(dt/d\tau)^2 - 1$

$= (m^*/ac^2)\,(1 + e\cos \mathrm{E})/(1 - e\cos \mathrm{E})$

Now neglecting c^{-4} with regard to 1, it follows that

$ndt = nd\tau\,[\,1 + \tfrac{1}{2}\,m^*/ac^2\,(1 + e\cos \mathrm{E})/(1 - e\cos \mathrm{E})\,]$

from which, by applying (29),

(31) $nt + \text{const} = (1 + \tfrac{1}{2}\,m^*/ac^2)\,n\tau + m^*/ac^2\,\mathrm{Sin}\,\mathrm{E}.$

the factor m^*/ac^2 is here the square of the ratio of a certain average velocity of F in its orbit to the velocity of light. If now m^* denote the mass of the sun, a the semi major axis of the earth's orbit, then this factor amounts to 10^{-8}.

The law of mass attraction which has been just described and which is formulated in accordance with the relativity postulate would signify that gravitation is propagated with the velocity of light. In view of the fact that the periodic terms in (31) are very small, it is not possible to decide out of astronomical observations between such a law (with the modified mechanics proposed above) and the Newtonian law of attraction with Newtonian mechanics.

SPACE AND TIME

A Lecture delivered before the Naturforscher Versammlung (Congress of Natural Philosophers) at Cologne—(21st September, 1908).

Gentlemen,

The conceptions about time and space, which I hope to develop before you to-day, has grown on experimental physical grounds. Herein lies its strength. The tendency is radical. Henceforth, the old conception of space for itself, and time for itself shall reduce to a mere shadow, and some sort of union of the two will be found consistent with facts.

I

Now I want to show you how we can arrive at the changed concepts about time and space from mechanics, as accepted now-a-days, from purely mathematical considerations. The equations of Newtonian mechanics show a twofold invariance, (*i*) their form remains unaltered when we subject the fundamental space-coordinate system to any possible change of position, (*ii*) when we change the system in its nature of motion, *i. e.*, when we impress upon it any uniform motion of translation, the null-point of time plays no part. We are accustomed to look upon the axioms of geometry as settled once for all, while we seldom have the same amount of conviction regarding the axioms of mechanics, and therefore the two invariants are seldom mentioned in the same breath. Each one of these denotes a certain group of transformations for the differential equations of mechanics. We look upon the existence of the first group as a fundamental characteristics of space. We always prefer to leave off the second group to itself, and with a light heart conclude that we can never decide from physical considerations whether the space, which is supposed to be at rest, may not finally be in uniform motion. So these two groups lead quite separate existences besides each other. Their totally heterogeneous character may scare us away from the attempt to compound them. Yet it is the whole compounded group which as a whole gives us occasion for thought.

We wish to picture to ourselves the whole relation graphically. Let (x, y, z) be the rectangular coordinates of space, and t denote the time. Subjects of our perception are always connected with place and time. *No one has observed a place except at a particular time, or has observed a time except at a particular place.* Yet I respect the dogma that time and space have independent existences. I will call a space-point plus a time-point, *i.e.*, a system of values x, y, z, t, as a *world-point.* The manifoldness of all possible values of x, y, z, t, will be the *world.* I can draw four world-axes with the chalk. Now any axis drawn consists of quickly vibrating molecules, and besides, takes part in all the journeys of the earth ; and therefore gives us occasion for reflection. The greater abstraction required for the four-axes does not cause the mathematician any trouble. In order not to allow any yawning gap to exist, we shall suppose that at every place and time, something perceptible exists. In order not to specify either matter or electricity, we shall simply style these as substances. We direct our attention to the *world-point* x, y, z, t, and suppose that we are in a position to recognise this substantial point at any subsequent time. Let dt be the time element corresponding to the changes of space coordinates of this point $[dx, dy, dz]$. Then we obtain (as a picture, so to speak, of the perennial life-career of the substantial point),—a curve in the *world*—the *world-line*, the points on which unambiguously correspond to the parameter t from $+\infty$ to $-\infty$. The whole world appears to be resolved in such *world-lines*, and I may just deviate from my point if I say that according to my opinion the physical laws would find their fullest expression as mutual relations among these lines.

By this conception of time and space, the (x, y, z) manifoldness $t = 0$ and its two sides $t < 0$ and $t > 0$ falls asunder. If for the sake of simplicity, we keep the null-point of time and space fixed, then the first named group of mechanics signifies that at $t = 0$ we can give the x, y, and z-axes any possible rotation about the null-point corresponding to the homogeneous linear transformation of the expression

$$x^2 + y^2 + z^2.$$

The second group denotes that without changing the expression for the mechanical laws, we can substitute $(x - \alpha t, y - \beta t, z - \gamma t$ for (x, y, z) where (α, β, γ) are any constants. According to this we can give the time-axis any possible direction in the upper half of the world $t > 0$. Now what have the demands of orthogonality in space to do with this perfect freedom of the time-axis towards the upper half?

To establish this connection, let us take a positive parameter c, and let us consider the figure

$$c^2t^2 - x^2 - y^2 - z^2 = 1$$

According to the analogy of the hyperboloid of two sheets, this consists of two sheets separated by $t = 0$. Let us consider the sheet, in the region of $t > 0$, and let us now conceive the transformation of x, y, z, t in the new system of variables; (x', y', z', t') by means of which the form of the expression will remain unaltered. Clearly the rotation of space round the null-point belongs to this group of transformations. Now we can have a full idea of the transformations which we picture to ourselves from a particular transformation in which (y, z) remain unaltered. Let us draw the cross section of the upper sheets with the plane of the x- and t-axes, *i.e.*, the upper half of the hyperbola $c^2t^2 - x^2 = 1$, with its asymptotes (*vide* fig. 1).

Then let us draw the radius rector OA', the tangent A' B' at A', and let us complete the parallelogram OA' B' C'; also produce B' C' to meet the x-axis at D'. Let us now take Ox', OA' as new axes with the unit measuring rods OC' = 1, OA' = (1/c) ; then the hyperbola is again expressed in the form $c^2t'^2 - x'^2 = 1$, $t' > 0$ and the transition from (x, y, z, t) to $(x' y' z' t)$ is one of the transitions in question. Let us add to this characteristic transformation any possible displacement of the space and time null-points; then we get a group of transformation depending only on c, which we may denote by G_c.

Now let us increase c to infinity. Thus $(1/c)$ becomes zero and it appears from the figure that the hyperbola is gradually shrunk into the x-axis, the asymptotic angle becomes a straight one, and every special transformation in the limit changes in such a manner that the t-axis can have any possible direction upwards, and x' more and more approximates to x. Remembering this point it is clear that the full group belonging to Newtonian Mechanics is simply the group G_c, with the value of $c = \infty$. In this state of affairs, and since G_c is mathematically more intelligible than G_∞, a mathematician may, by a free play of imagination, hit upon the thought that natural phenomena possess an invariance not only for the group G_∞, but in fact also for a group G_c, where c is finite, but yet exceedingly large compared to the usual measuring units. Such a preconception would be an extraordinary triumph for pure mathematics.

At the same time I shall remark for which value of c, this invariance can be conclusively held to be true. *For c, we shall substitute the velocity of light c in free space.* In order to avoid speaking either of space or of vacuum, we may take this quantity as the ratio between the electrostatic and electro-magnetic units of electricity.

We can form an idea of the invariant character of the expression for natural laws for the group-transformation G_c in the following manner.

Out of the totality of natural phenomena, we can, by successive higher approximations, deduce a coordinate system (x, y, z, t); by means of this coordinate system, we can represent the phenomena according to definite laws. This system of reference is by no means uniquely determined by the phenomena. *We can change the system of reference in any possible manner corresponding to the above-mentioned group transformation G_c, but the expressions for natural laws will not be changed thereby.*

For example, corresponding to the above described figure, we can call t' the time, but then necessarily the space connected with it must be expressed by the manifoldness $(x' y z)$. The physical laws are now expressed by means of x', y, z, t',—and the expressions are just the same as in the case of x, y, z, t. According to this, we shall have in the world, not one space, but many spaces,—quite analogous to the case that the three-dimensional space consists of an infinite number of planes. The three-dimensional geometry will be a chapter of four-dimensional physics. Now you perceive, why I said in the beginning that time and space shall reduce to mere shadows and we shall have a world complete in itself.

II

Now the question may be asked,—what circumstances lead us to these changed views about time and space, are they not in contradiction with observed phenomena, do they finally guarantee us advantages for the description of natural phenomena?

Before we enter into the discussion, a very important point must be noticed. Suppose we have individualised time and space in any manner; then a world-line parallel to the t-axis will correspond to a stationary point; a world-line inclined to the t-axis will correspond to a point moving uniformly; and a world-curve will correspond to a point moving in any manner. Let us now picture to our mind the world-line passing through any world point x, y, z, t; now if we find the world-line parallel to the radius vector OA' of the hyperboloidal sheet, then we can introduce OA' as a new time-axis, and then according to the new conceptions of time and space the substance will appear to be at rest in the world point concerned. We shall now introduce this fundamental axiom:—

The substance existing at any world point can always be conceived to be at rest, if we establish our time and space suitably. The axiom denotes that in a world-point, the expression

$$c^2dt^2 - dx^2 - dy^2 - dz^2$$

shall always be positive or what is equivalent to the same thing, every velocity V should be smaller than c. c shall therefore be the upper limit for all substantial velocities and herein lies a deep significance for the quantity c. At the first impression, the axiom seems to be rather unsatisfactory. It is to be remembered that only a modified mechanics will occur, in which the square root of this differential combination takes the place of time, so that cases in which the velocity is greater than c will play no part, something like imaginary coordinates in geometry.

The *impulse* and real cause of inducement *for the assumption of the group-transformation* G_c is the fact that the differential equation for the propagation of light in vacant space possesses the group-transformation G_c. On the other hand, the idea of rigid bodies has any sense only in a system mechanics with the group G_{infinity}. Now if we have an optics with G_c, and on the other hand if there are rigid bodies, it is easy to see that a t-direction can be defined by the two hyperboloidal shells common to the groups G_∞, and G_c, which has got the further consequence, that by means of suitable rigid instruments in the laboratory, we can perceive a change in natural phenomena, in case of different orientations, with regard to the direction of progressive motion of the earth. But all efforts directed towards this object, and even the celebrated interference-experiment of Michelson have given negative results. In order to supply an explanation for this result, H. A. Lorentz formed a hypothesis which practically amounts to an invariance of optics for the group G_c. According to Lorentz every substance shall suffer a contraction

$1:(\sqrt{(1 - v^2/c^2)})$ in length, in the direction of its motion

$l/l' = 1/\sqrt{(1 - v^2/c^2)}$ $l' = l(1 - v^2/c^2)$.

This hypothesis sounds rather phantastical. For the contraction is not to be thought of as a consequence of the resistance of ether, but purely as a gift from the skies, as a sort of condition always accompanying a state of motion.

I shall show in our figure that Lorentz's hypothesis is fully equivalent to the new conceptions about time and space. Thereby it may appear more intelligible. Let us now, for the sake of simplicity, neglect (y, z) and fix our attention on a two dimensional world, in which let upright strips parallel to the t-axis represent a state of rest and another parallel strip inclined to the t-axis represent a state of uniform motion for a body, which has a constant spatial extension (see fig. 1). If OA' is parallel to the second strip, we can take t' as the t-axis and x' as the x-axis, then the second body will appear to be at rest, and the first body in uniform motion. We shall now assume that the first body supposed to be at rest, has the length l, *i.e.*, the cross section PP of the first strip upon the x-axis = $l \cdot$ OC, where OC is the unit measuring rod upon the x-axis—and the second body also, when supposed to be at rest, has the same length l, this means that, the cross section Q'Q' of the second strip has a cross-section $l \cdot$ OC', when measured parallel to the x'-axis. In these two bodies, we have now images of two Lorentz-electrons, one of which is at rest and the other moves uniformly. Now if we stick to our original coordinates, then the extension of the second electron is given by the cross section QQ of the strip belonging to it measured parallel to the x-axis. Now it is clear since Q'Q' = $l \cdot$ OC', that QQ = $l \cdot$ OD'.

If $(dc/dt) = v$, an easy calculation gives that

OD' = OC $\sqrt{(1-(v^2/c^2))}$, therefore (PP/QQ) = $(1/\sqrt{(1-(v^2/c^2)})$

This is the sense of Lorentz's hypothesis about the contraction of electrons in case of motion. On the other hand, if we conceive the second electron to be at rest, and therefore adopt the system $(x', t',)$ then the cross-section P'P' of the strip of the electron parallel to OC' is to be regarded as its length and we shall find the first electron shortened with reference to the second in the same proportion, for it is,

P'P'/Q'Q' = OD/OC' = OD'/OC = QQ/PP

Lorentz called the combination t' of (t and x) as the *local time* (*Ortszeit*) of the uniformly moving electron, and used a physical construction of this idea for a better comprehension of the contraction-hypothesis. But to perceive clearly that the time of an electron is as good as the time of any other electron, *i.e.* t, t' are to be regarded as equivalent, has been the service of A. Einstein [Ann. d. Phys. 891, p. 1905, Jahrb. d. Radis. ... 4-4-11-1907.] There the concept of time was shown to be completely and unambiguously established by natural phenomena. But the concept of space was not arrived at, either by Einstein or Lorentz, probably because in the case of the above-mentioned spatial transformations, where the (x', t') plane coincides with the x-t plane, the significance is possible that the x-axis of space somehow remains conserved in its position.

We can approach the idea of space in a corresponding manner, though some may regard the attempt as rather fantastical.

According to these ideas, the word "Relativity-Postulate" which has been coined for the demands of invariance in the group G, seems to be rather inexpressive for a true understanding of the group G_c, and for further progress. Because the sense of the postulate is that the four-dimensional world is given in space and time by phenomena only, but the projection in time and space can be handled with a certain freedom, and therefore I would rather like to give to this assertion the name "*The Postulate of the Absolute world*" [World-Postulate].

III

By the world-postulate a similar treatment of the four determining quantities x, y, z, t, of a world-point is possible. Thereby the forms under which the physical laws come forth, gain in intelligibility, as I shall presently show. Above all, the idea of acceleration becomes much more striking and clear.

I shall again use the geometrical method of expression. Let us call any world-point O as a "Space-time-null-point." The cone

$$c^2t^2 - x^2 - y^2 - z^2 = O$$

consists of two parts with O as apex, one part having $t < 0$, the other having $t > 0$. The first, which we may call *the fore-cone* consists of all those points which send light towards O, the second, which we may call *the aft-cone,* consists of all those points which receive their light from O. The region bounded by the fore-cone may be called the fore-side of O, and the region bounded by the aft-cone may be called the aft-side of O. (*Vide* fig. 2).

On the aft-side of O we have the already considered hyperboloidal shell F $= c^2t^2 - x^2 - y^2 - z^2 = 1, t > 0$.

The region inside the two cones will be occupied by the hyperboloid of one sheet

$$-F = x^2 + y^2 + z^2 - c^2t^2 = k^2,$$

where k^2 can have all possible positive values. The hyperbolas which lie upon this figure with O as centre, are important for us. For the sake of clearness the individual branches of this hyperbola will be called the "*Inter-hyperbola with centre O.*" Such a hyperbolic branch, when thought of as a world-line, would represent a motion which for $t = -\infty$ and $t = \infty$, asymptotically approaches the velocity of light c.

If, by way of analogy to the idea of vectors in space, we call any directed length in the manifoldness x, y, z, t a vector, then we have to distinguish between a time-vector directed from O towards the sheet $\pm F = 1, t > 0$ and a space-vector directed from O towards the sheet $-F = 1$. The time-axis can be parallel to any vector of the first kind. Any world-point between the *fore* and *aft cones* of O, may by means of the system of reference be regarded either as synchronous with O, as well as later or earlier than O. Every world-point on the fore-side of O is necessarily always earlier, every point on the aft side of O, later than O. The limit $c = \infty$ corresponds to a complete folding up of the wedge-shaped cross-section between the fore and aft cones in the manifoldness $t = 0$. In the figure drawn, this cross-section has been intentionally drawn with a different breadth.

Let us decompose a vector drawn from O towards (x, y, z, t) into its components. If the directions of the two vectors are respectively the directions of the radius vector OR to one of the surfaces $\pm F = 1$, and of a tangent RS at the point R of the surface, then the vectors shall be called normal to each other. Accordingly

$$c^2tt_1 - xx_1 - yy_1 - zz_1 = 0,$$

which is the condition that the vectors with the components (x, y, z, t) and $(x_1\, y_1\, z_1\, t_1)$ are normal to each other.

For the *measurement* of vectors in different directions, the unit measuring rod is to be fixed in the following manner;—a space-like vector from 0 to $-F = I$ is always to have the measure unity, and a time-like vector from O to $+F = 1$, $t > 0$ is always to have the measure $1/c$.

Let us now fix our attention upon the world-line of a substantive point running through the world-point (x, y, z, t); then as we follow the *progress* of the line, the quantity

$$d\tau = (1/c)\sqrt{(c^2dt^2 - dx^2 - dy^2 - dz^2)},$$

corresponds to the time-like vector-element (dx, dy, dz, dt).

The integral $\tau = \int d\tau$, taken over the world-line from any fixed initial point P_0 to any variable final point P, may be called the "Proper-time" of the substantial point at P_0 upon the *world-line*. We may regard (x, y, z, t), *i.e.*, the components of the vector OP, as functions of the "proper-time" τ; let ([.x], [.y], [.z], [.t]) denote the first differential-quotients, and ([..x], [..y], [..z], [..t]) the second differential quotients of (x, y, z, t) with regard to τ, then these may respectively be called the *Velocity-vector*, and the *Acceleration-vector* of the substantial point at P. Now we have

$$c^2\,[.t^2] - [.x^2] - [.y^2] - [.z^2] = c^2$$

$$c^2\,[.t][..t] - [.x][..x] - [.y][..y] - [.z][..z] = 0$$

i.e., the '*Velocity-vector*' is the time-like vector of unit measure in the direction of the world-line at P, the '*Acceleration-vector*' at P is normal to the velocity-vector at P, and is in any case, a space-like vector.

Now there is, as can be easily seen, a certain hyperbola, which has three infinitely contiguous points in common with the world-line at P, and of which the asymptotes are the generators of a 'fore-cone' and an 'aft-cone.' This hyperbola may be called the "hyperbola of curvature" at P (*vide* fig. 3). If M be the centre of this hyperbola, then we have to deal here with an 'Inter-hyperbola' with centre M. Let P = measure of the vector MP, then we easily perceive that the acceleration-vector at P is *a vector of magnitude* c^2/ϱ *in the direction of* MP.

If $[..x]$, $[..y]$, $[..z]$, $[..t]$ are nil, then the hyperbola of curvature at P reduces to the straight line touching the world-line at P, and $\varrho = \infty$.

IV

In order to demonstrate that the assumption of the group G_c for the physical laws does not possibly lead to any contradiction, it is unnecessary to undertake a revision of the whole of physics on the basis of the assumptions underlying this group. The revision has already been successfully made in the case of "Thermodynamics and Radiation,"[30] for "Electromagnetic phenomena",[31] and finally for "Mechanics with the maintenance of the idea of mass."

For this last mentioned province of physics, the question may be asked: if there is a force with the components X, Y, Z (in the direction of the space-axes) at a world-point (x, y, z, t), where the velocity-vector is ([.x], [.y], [.z], [.t]), then how are we to regard this force when the system of reference is changed in any possible manner? Now it is known that there are certain well-tested theorems about the ponderomotive force in electromagnetic fields, where the group G_c is undoubtedly permissible. These theorems lead us to the following simple rule; *if the system of reference be changed in any way, then the supposed force is to be put as a force in the new space-coordinates in such a manner, that the corresponding vector with the components*

[.t]X, [.t]Y, [.t]Z, [.t]T,

where T = $1/c^2$ ([.x]/[.t] X + [.y]/[.t] Y + [.z]/[.t] Z) = $1/c^2$

(*the rate of*

which work is done at the world-point), *remains unaltered.*

This vector is always normal to the velocity-vector at P. Such a force-vector, representing a force at P, may be called a *moving force-vector at* P.

Now the world-line passing through P will be described by a substantial point with the constant *mechanical mass m*. Let us call *m-times* the velocity-vector at P as the *impulse-vector*, and *m-times* the acceleration-vector at P as the *force-vector of motion*, at P. According to these definitions, the following law tells us how the motion of a point-mass takes place under any moving force-vector[32]:

The force-vector of motion is equal to the moving force-vector.

This enunciation comprises four equations for the components in the four directions, of which the fourth can be deduced from the first three, because both of the above-mentioned vectors are perpendicular to the velocity-vector. From the definition of T, we see that the fourth simply expresses the "Energy-law." Accordingly *c^2-times the component of the impulse-vector in the* *direction of the t-axis is* to be defined as *the kinetic-energy* of the point-mass. The expression for this is

$$mc^2\, dt/d\tau = mc^2 / \sqrt{(1 - v^2/c^2)}$$

i.e., if we deduct from this the additive constant mc^2, we obtain the expression $\frac{1}{2}\, mv^2$ of Newtonian-mechanics up to magnitudes of *the order of* $1/c^2$. Hence it appears that *the energy* depends *upon the system of reference.* But since the t-axis can be laid in the direction of any time-like axis, therefore the energy-law comprises, for any possible system of reference, the whole system of equations of motion. This fact retains its significance even in the limiting case $c = \infty$, for the axiomatic construction of Newtonian mechanics, as has already been pointed out by T. R. Schütz.[33]

From the very beginning, we can establish the ratio between the units of time and space in such a manner, that the velocity of light becomes unity. If we now write $\sqrt{-1}\; t = l$, in the place of l, then the differential expression

$$d\tau^2 = -(dx^2 + dy^2 + dz^2 + dl^2),$$

becomes symmetrical in (x, y, r, l); this symmetry then enters into each law, which does not contradict the *world-postulate.* We can clothe the "essential nature of this postulate in the mystical, but mathematically significant formula

$$3 \cdot 10^5\; km = \sqrt{-1}\; \text{Sec.}$$

V

The advantages arising from the formulation of the world-postulate are illustrated by nothing so strikingly as by the expressions which tell us about the reactions exerted by a point-charge moving in any manner according to the Maxwell-Lorentz theory.

Let us conceive of the world-line of such an electron with the charge (e), and let us introduce upon it the "Proper-time" τ reckoned from any possible initial point. In order to obtain the field caused by the electron at any world-point P_1 let us construct the fore-cone belonging to P_1 (*vide* fig. 4). Clearly this cuts the unlimited world-line of the electron at a single point P, because these directions are all time-like vectors. At P, let us draw the tangent to the world-line, and let us draw from P_1 the normal to this tangent. Let r be the measure of P_1Q. According to the definition of a fore-cone, r/e is to be reckoned as the measure of PQ. Now at the world-point P_1, the vector-potential of the field excited by e is represented by the vector in direction PQ, having the magnitude e/cr, in its three space components along the x-, y-, z-axes; the scalar-potential is represented by the component along the t-axis. This is the elementary law found out by A. Lienard, and E. Wiechert.[34]

If the field caused by the electron be described in the above-mentioned way, then it will appear that the division of the field into electric and magnetic forces is a relative one, and depends upon the time-axis assumed; the two forces considered together bears some analogy to the force-screw in mechanics; the analogy is, however, imperfect.

I shall now describe *the ponderomotive force which is exerted by one moving electron upon another moving electron.* Let us suppose that the world-line of a second point-electron passes through the world-point P_1. Let us determine P, Q, r as before, construct the middle-point M of the hyperbola of curvature at P, and finally the normal MN upon a line through P which is parallel to QP_1. With P as the initial point, we shall establish a system of reference in the following way: the t-axis will be laid along PQ, the x-axis in the direction of QP_1. The y-axis in the direction of MN, then the z-axis is automatically determined, as it is normal to the x-, y-, z-axes. Let $[\ddot{x}]$, $[\ddot{y}]$, $[\ddot{z}]$, $[\ddot{t}]$ be the acceleration-vector at P, $[\dot{x}]_1$, $[\dot{y}]_1$ $[\dot{z}]_1$, $[\dot{t}]_1$ be the velocity-vector at P_1. Then the force-vector exerted by the first election e, (moving in any possible manner) upon the second election e, (likewise moving in any possible manner) at P_1 is represented by

$$-e\, e_1([\dot{t}_1] - [\dot{x}_1]/c)F,$$

For the components F_x, F_y, F_z, F_t *of the vector* F *the following three relations hold:—*

$$cF_t - F_x = 1/r^2,\ F_y = [\ddot{y}]/(c^2r),\ F_z = 0,$$

and fourthly this vector F *is normal to the velocity-vector* P_1, *and through this circumstance alone, its dependence on this last velocity-vector arises.*

If we compare with this expression the previous formulæ[35] giving the elementary law about the ponderomotive action of moving electric charges upon each other, then we cannot but admit, that the relations which occur here reveal the inner essence of full simplicity first in four dimensions; but in three dimensions, they have very complicated projections.

In the mechanics reformed according to the world-postulate, the disharmonies which have disturbed the relations between Newtonian mechanics and modern electrodynamics automatically disappear. I shall now consider the position of the Newtonian law of attraction to this postulate. I will assume that two point-masses m and m_1 describe their world-lines; a moving force-vector is exercised by m upon m_1, and the expression is just the same as in the case of the electron, only we have to write $+mm_1$ instead $-ee_1$. We shall consider only the special case in which the acceleration-vector of m is always zero: then t may be introduced in such a manner that m may be regarded as fixed, the motion of m is now subjected to the moving-force vector of m alone. If we now modify this given vector by writing $-([\cdot]1/\sqrt{(1-(v^2/c^2))}$ instead of $[.t]$ ($[.t] = 1$ up to magnitudes of the order $(1[\cdot]/c^2)$), then it appears that Kepler's laws hold good for the position (x_1, y_1, z_1), of m_1 at any time, only in place of the time t_1, we have to write the proper time τ_1 of m_1. On the basis of this simple remark, it can be seen that the proposed law of attraction in combination with new mechanics is not less suited for the explanation of astronomical phenomena than the Newtonian law of attraction in combination with Newtonian mechanics.

Also the fundamental equations for electro-magnetic processes in moving bodies are in accordance with the world-postulate. I shall also show on a later occasion that the deduction of these equations, as taught by Lorentz, are by no means to be given up.

The fact that the world-postulate holds without exception is, I believe, the true essence of an electromagnetic picture of the world; the idea first occurred to Lorentz, its essence was first picked out by Einstein, and is now gradually fully manifest. In course of time, the mathematical consequences will be gradually deduced, and enough suggestions will be forthcoming for the experimental verification of the postulate; in this way even those, who find it uncongenial, or even painful to give up the old, time-honoured concepts, will be reconciled to the new ideas of time and space,—in the prospect that they will lead to pre-established harmony between pure mathematics and physics.

The Foundation of the Generalised Theory of Relativity By A. Einstein. From Annalen der Physik 4.49.1916.

The theory which is sketched in the following pages forms the most wide-going generalization conceivable of what is at present known as "the theory of Relativity;" this latter theory I differentiate from the former "Special Relativity theory," and suppose it to be known. The generalization of the Relativity theory has been made much easier through the form given to the special Relativity theory by Minkowski, which mathematician was the first to recognize clearly the formal equivalence of the space like and time-like co-ordinates, and who made use of it in the building up of the theory. The mathematical apparatus useful for the general relativity theory, lay already complete in the "Absolute Differential Calculus," which were based on the researches of Gauss, Riemann and Christoffel on the non-Euclidean manifold, and which have been shaped into a system by Ricci and Levi-civita, and already applied to the problems of theoretical physics. I have in part B of this communication developed in the simplest and clearest manner, all the supposed mathematical auxiliaries, not known to Physicists, which will be useful for our purpose, so that, a study of the mathematical literature is not necessary for an understanding of this paper. Finally in this place I thank my friend Grossmann, by whose help I was not only spared the study of the mathematical literature pertinent to this subject, but who also aided me in the researches on the field equations of gravitation.

A

Principal considerations about the Postulate of Relativity.

§ 1. Remarks on the Special Relativity Theory.

The special relativity theory rests on the following postulate which also holds valid for the Galileo-Newtonian mechanics.

If a co-ordinate system K be so chosen that when referred to it, the physical laws hold in their simplest forms these laws would be also valid when referred to another system of co-ordinates K' which is subjected to an uniform translational motion relative to K. We call this postulate "The Special Relativity Principle." By the word special, it is signified that the principle is limited to the case, when K' has *uniform translatory* motion with reference to K, but the equivalence of K and K' does not extend to the case of non-uniform motion of K' relative to K.

The Special Relativity Theory does not differ from the classical mechanics through the assumption of this postulate, but only through the postulate of the constancy of light-velocity in vacuum which, when combined with the special relativity postulate, gives in a well-known way, the relativity of synchronism as well as the Lorenz-transformation, with all the relations between moving rigid bodies and clocks.

The modification which the theory of space and time has undergone through the special relativity theory, is indeed a profound one, but a weightier point remains untouched. According to the special relativity theory, the theorems of geometry are to be looked upon as the laws about any possible relative positions of solid bodies at rest, and more generally the theorems of kinematics, as theorems which describe the relation between measurable bodies and clocks. Consider two material points of a solid body at rest; then according to these conceptions there corresponds to these points a wholly definite extent of length, independent of kind, position, orientation and time of the body.

Similarly let us consider two positions of the pointers of a clock which is at rest with reference to a co-ordinate system; then to these positions, there always corresponds, a time-interval of a definite length, independent of time and place. It would be soon shown that the general relativity theory can not hold fast to this simple physical significance of space and time.

§ 2. About the reasons which explain the extension of the relativity-postulate.

To the classical mechanics (no less than) to the special relativity theory, is attached an episteomological defect, which was perhaps first cleanly pointed out by E. Mach. We shall illustrate it by the following example; Let two fluid bodies of equal kind and magnitude swim freely in space at such a great distance from one another (and from all other masses) that only that sort of gravitational forces are to be taken into account which the parts of any of these bodies exert upon each other. The distance of the bodies from one another is invariable. The relative motion of the different parts of each body is not to occur. But each mass is seen to rotate by an observer at rest relative to the other mass round the connecting line of the masses with a constant angular velocity (definite relative motion for both the masses). Now let us think that the surfaces of both the bodies (S_1 and S_2) are measured with the help of measuring rods (relatively at rest); it is then found that the surface of S_1 is a sphere and the surface of the other is an ellipsoid of rotation. We now ask, why is this difference between the two bodies? An answer to this question can only then be regarded as satisfactory from the episteomological standpoint when the thing adduced as the cause is an observable fact of experience. The law of causality has the sense of a definite statement about the world of experience only when observable facts alone appear as causes and effects.

The Newtonian mechanics does not give to this question any satisfactory answer. For example, it says:—The laws of mechanics hold true for a space R_1 relative to which the body S_1 is at rest, not however for a space relative to which S_2 is at rest.

The Galiliean space, which is here introduced is however only a purely imaginary cause, not an observable thing. It is thus clear that the Newtonian mechanics does not, in the case treated here, actually fulfil the requirements of causality, but produces on the mind a fictitious complacency, in that it makes responsible a *wholly imaginary cause* R_1 for the different behaviours of the bodies S_1 and S_2 which are actually observable.

A satisfactory explanation to the question put forward above can only be thus given:—that the physical system composed of S_1 and S_2 shows for itself alone no conceivable cause to which the different behaviour of S_1 and S_2 can be attributed. The cause must thus lie outside the system. We are therefore led to the conception that the general laws of motion which determine specially the forms of S_1 and S_2 must be of such a kind, that the mechanical behaviour of S_1 and S_2 must be essentially conditioned by the distant masses, which we had not brought into the system considered. These distant masses, (and their relative motion as regards the bodies under consideration) are then to be looked upon as the seat of the principal observable causes for the different behaviours of the bodies under consideration. They take the place of the imaginary cause R_1. Among all the conceivable spaces R_1 and R_2 moving in any manner relative to one another, there is a priori, no one set which can be regarded as affording greater advantages, against which the objection which was already raised from the standpoint of the theory of knowledge cannot be again revived. The laws of physics must be so constituted that they should remain valid for any system of co-ordinates moving in any manner. We thus arrive at an extension of the relativity postulate.

Besides this momentous episteomological argument, there is also a well-known physical fact which speaks in favour of an extension of the relativity theory. Let there be a Galiliean co-ordinate system K relative to which (at least in the four-dimensional region considered) a mass at a sufficient distance from other masses move uniformly in a line. Let K' be a second co-ordinate system which has a uniformly accelerated motion relative to K. Relative to K' any mass at a sufficiently great distance experiences an accelerated motion such that its acceleration and the direction of acceleration is independent of its material composition and its physical conditions.

Can any observer, at rest relative to K', then conclude that he is in an actually accelerated reference-system? This is to be answered in the negative; the above-named behaviour of the freely moving masses relative to K' can be explained in as good a manner in the following way. The reference-system K' has no acceleration. In the space-time region considered there is a gravitation-field which generates the accelerated motion relative to K'.

This conception is feasible, because to us the experience of the existence of a field of force (namely the gravitation field) has shown that it possesses the remarkable property of imparting the same acceleration to all bodies. The mechanical behaviour of the bodies relative to K' is the same as experience would expect of them with reference to systems which we assume from habit as stationary; thus it explains why from the physical stand-point it can be assumed that the systems K and K' can both with the same legitimacy be taken as at rest, that is, they will be equivalent as systems of reference for a description of physical phenomena.

From these discussions we see, that the working out of the general relativity theory must, at the same time, lead to a theory of gravitation; for we can "create" a gravitational field by a simple variation of the co-ordinate system. Also we see immediately that the principle of the constancy of light-velocity must be modified, for we recognise easily that the path of a ray of light with reference to K' must be, in general, curved, when light travels with a definite and constant velocity in a straight line with reference to K.

§ 3. The time-space continuum. Requirements of the general Co-variance for the equations expressing the laws of Nature in general.

In the classical mechanics as well as in the special relativity theory, the co-ordinates of time and space have an immediate physical significance; when we say that any arbitrary point has x_1 as its X_1 co-ordinate, it signifies that the projection of the point-event on the X_1-axis *ascertained* by means of a solid rod according to the rules of Euclidean Geometry is reached when a definite measuring rod, the unit rod, can be carried x_1 times from the origin of co-ordinates along the X_1 axis. A point having $x_4 = t_1$ as the X_4 co-ordinate signifies that a unit clock which is adjusted to be at rest relative to the system of co-ordinates, and coinciding in its spatial position with the point-event and set according to some definite standard has gone over $x_4 = t$ periods before the occurrence of the point-event.

This conception of time and space is continually present in the mind of the physicist, though often in an unconscious way, as is clearly recognised from the role which this conception has played in physical measurements. This conception must also appear to the reader to be lying at the basis of the second consideration of the last paragraph and imparting a sense to these conceptions. But we wish to show that we are to abandon it and in general to replace it by more general conceptions in order to be able to work out thoroughly the postulate of general relativity,—the case of special relativity appearing as a limiting case when there is no gravitation.

We introduce in a space, which is free from Gravitation-field, a Galiliean Co-ordinate System K (x, y, z, t) and also, another system K' $(x' y' z' t')$ rotating uniformly relative to K. The origin of both the systems as well as their z-axes might continue to coincide. We will show that for a space-time measurement in the system K', the above established rules for the physical significance of time and space can not be maintained. On grounds of symmetry it is clear that a circle round the origin in the XY plane of K, can also be looked upon as a circle in the plane (X', Y') of K'. Let us now think of measuring the circumference and the diameter of these circles, with a unit measuring rod (infinitely small compared with the radius) and take the quotient of both the results of measurement. If this experiment be carried out with a measuring rod at rest relatively to the Galiliean system K we would get π, as the quotient. The result of measurement with a rod relatively at rest as regards K' would be a number which is greater than π. This can be seen easily when we regard the whole measurement-process from the system K and remember that the rod placed on the periphery suffers a Lorenz-contraction, not however when the rod is placed along the radius. Euclidean Geometry therefore does not hold for the system K'; the above fixed conceptions of co-ordinates which assume the validity of Euclidean Geometry fail with regard to the system K'. We cannot similarly introduce in K' a time corresponding to physical requirements, which will be shown by all similarly prepared clocks at rest relative to the system K'. In order to see this we suppose that two similarly made clocks are arranged one at the centre and one at the periphery of the circle, and considered from the stationary system K. According to the well-known results of the special relativity theory it follows—(as viewed from K)—that the clock placed at the periphery will go slower than the second one which is at rest. The observer at the common origin of co-ordinates who is able to see the clock at the periphery by means of light will see the clock at the periphery going slower than the clock beside him. Since he cannot allow the velocity of light to depend explicitly upon the time in the way under consideration he will interpret his observation by saying that the clock on the periphery actually goes slower than the clock at the origin. He cannot therefore do otherwise than define time in such a way that the rate of going of a clock depends on its position.

We therefore arrive at this result. In the general relativity theory time and space magnitudes cannot be so defined that the difference in spatial co-ordinates can be immediately measured by the unit-measuring rod, and time-like co-ordinate difference with the aid of a normal clock.

The means hitherto at our disposal, for placing our co-ordinate system in the time-space continuum, in a definite way, therefore completely fail and it appears that there is no other way which will enable us to fit the co-ordinate system to the four-dimensional world in such a way, that by it we can expect to get a specially simple formulation of the laws of Nature. So that nothing remains for us but to regard all conceivable co-ordinate systems as equally suitable for the description of natural phenomena. This amounts to the following law:—

That in general, Laws of Nature are expressed by means of equations which are valid for all co-ordinate systems, that is, which are covariant for all possible transformations. It is clear that a physics which satisfies this postulate will be unobjectionable from the standpoint of the general relativity postulate. Because among all substitutions there are, in every case, contained those, which correspond to all relative motions of the co-ordinate system (in three dimensions). This condition of general covariance which takes away the last remnants of physical objectivity from space and time, is a natural requirement, as seen from the following considerations. All our *well-substantiated* space-time propositions amount to the determination of space-time coincidences. If, for example, the event consisted in the motion of material points, then, for this last case, nothing else are really observable except the encounters between two or more of these material points. The results of our measurements are nothing else than well-proved theorems about such coincidences of material points, of our measuring rods with other material points, coincidences between the hands of a clock, dial-marks and point-events occurring at the same position and at the same time.

The introduction of a system of co-ordinates serves no other purpose than an easy description of totality of such coincidences. We fit to the world our space-time variables (x_1 x_2 x_3 x_4) such that to any and every point-event corresponds a system of values of (x_1 x_2 x_3 x_4). Two coincident point-events correspond to the same value of the variables (x_1 x_2 x_3 x_4); *i.e.*, the coincidence is characterised by the equality of the co-ordinates. If we now introduce any four functions (x'_1 x'_2 x'_3 x'_4) as co-ordinates, so that there is an unique correspondence between them, the equality of all the four co-ordinates in the new system will still be the expression of the space-time coincidence of two material points. As the purpose of all physical laws is to allow us to remember such coincidences, there is a priori no reason present, to prefer a certain co-ordinate system to another; *i.e.*, we get the condition of general covariance.

§ 4. Relation of four co-ordinates to spatial and time-like measurements.

Analytical expression for the Gravitation-field.

I am not trying in this communication to deduce the general Relativity-theory as the simplest logical system possible, with a minimum of axioms. But it is my chief aim to develop the theory in such a manner that the reader perceives the psychological naturalness of the way proposed, and the fundamental assumptions appear to be most reasonable according to the light of experience. In this sense, we shall now introduce the following supposition; that for an infinitely small four-dimensional region, the relativity theory is valid in the special sense when the axes are suitably chosen.

The nature of acceleration of an infinitely small (positional) co-ordinate system is hereby to be so chosen, that the gravitational field does not appear; this is possible for an infinitely small region. X_1, X_2, X_3 are the spatial co-ordinates; X_4 is the corresponding time-co-ordinate measured by some suitable measuring clock. These co-ordinates have, with a given orientation of the system, an immediate physical significance in the sense of the special relativity theory (when we take a rigid rod as our unit of measure). The expression

(1) $ds^2 = - dX_1^2 - dX_2^2 - dX_3^2 + dX_4^2$

had then, according to the special relativity theory, a value which may be obtained by space-time measurement, and which is independent of the orientation of the local co-ordinate system. Let us take ds as the magnitude of the line-element belonging to two infinitely near points in the four-dimensional region. If ds^2 belonging to the element (dX_1, dX_2, dX_3, dX_4) be positive we call it with Minkowski, time-like, and in the contrary case space-like.

To the line-element considered, *i.e.*, to both the infinitely near point-events belong also definite differentials dx_1, dx_2, dx_3, dx_4, of the four-dimensional co-ordinates of any chosen system of reference. If there be also a local system of the above kind given for the case under consideration, dX's would then be represented by definite linear homogeneous expressions of the form

(2) $dX_\nu = \sigma_\sigma a_{\nu\sigma} dx_\sigma$

If we substitute the expression in (1) we get

(3) $ds^2 = \sigma_{\sigma\tau} g_{\sigma\tau} dx_\sigma dx_\tau$

where $g_{\sigma\tau}$ will be functions of x_σ, but will no longer depend upon the orientation and motion of the 'local' co-ordinates; for ds^2 is a definite magnitude belonging to two point-events infinitely near in space and time and can be got by measurements with rods and clocks. The $g_{\tau\sigma}$'s are here to be so chosen, that $g_{\tau\sigma} = g_{\sigma\tau}$; the summation is to be extended over all values of σ and τ, so that the sum is to be extended, over 4×4 terms, of which 12 are equal in pairs.

From the method adopted here, the case of the usual relativity theory comes out when owing to the special behaviour of $g_{\sigma\tau}$ in a *finite* region it is possible to choose the system of co-ordinates in such a way that $g_{\sigma\tau}$ assumes constant values—

{ -1, 0, 0, 0

(4) { 0 -1 0 0

{ 0 0 -1 0

{ 0 0 0 +1

We would afterwards see that the choice of such a system of co-ordinates for a finite region is in general not possible.

From the considerations in § 2 and § 3 it is clear, that from the physical stand-point the quantities $g_{\sigma\tau}$ are to be looked upon as magnitudes which describe the gravitation-field with reference to the chosen system of axes. We assume firstly, that in a certain four-dimensional region considered, the special relativity theory is true for some particular choice of co-ordinates. The $g_{\sigma\tau}$'s then have the values given in (4). A free material point moves with reference to such a system uniformly in a straight-line. If we now introduce, by any substitution, the space-time co-ordinates $x_1 ... x_4$ then in the new system $g_{\mu\nu}$'s are no longer constants, but functions of space and time. At the same time, the motion of a free point-mass in the new co-ordinates, will appear as curvilinear, and not uniform, in which the law of motion, will be *independent of the nature of the moving mass-points.* We can thus signify this motion as one under the influence of a gravitation field. We see that the appearance of a gravitation-field is connected with space-time variability of $g_{\sigma\tau}$'s. In the general case, we can not by any suitable choice of axes, make special relativity theory valid throughout any finite region. We thus deduce the conception that $g_{\sigma\tau}$'s describe the gravitational field. According to the general relativity theory, gravitation thus plays an exceptional rôle as distinguished from the others, specially the electromagnetic forces, in as much as the 10 functions $g_{\sigma\tau}$ representing gravitation, define immediately the metrical properties of the four-dimensional region.

B

Mathematical Auxiliaries for Establishing the General Covariant Equations.

We have seen before that the general relativity-postulate leads to the condition that the system of equations for Physics, must be co-variants for any possible substitution of co-ordinates $x_1, \ldots x_4$; we have now to see how such general co-variant equations can be obtained. We shall now turn our attention to these purely mathematical propositions. It will be shown that in the solution, the invariant *ds*, given in equation (3) plays a fundamental rôle, which we, following Gauss's Theory of Surfaces, style as the line-element.

The fundamental idea of the general co-variant theory is this:—With reference to any co-ordinate system, let certain things (tensors) be defined by a number of functions of co-ordinates which are called the components of the tensor. There are now certain rules according to which the components can be calculated in a new system of co-ordinates, when these are known for the original system, and when the transformation connecting the two systems is known. The things herefrom designated as "Tensors" have further the property that the transformation equation of their components are linear and homogeneous; so that all the components in the new system vanish if they are all zero in the original system. Thus a law of Nature can be formulated by putting all the components of a tensor equal to zero so that it is a general co-variant equation; thus while we seek the laws of formation of the tensors, we also reach the means of establishing general co-variant laws.

5. Contra-variant and co-variant Four-vector.

Contra-variant Four-vector. The line-element is defined by the four components dx_ν, whose transformation law is expressed by the equation

$$dx'_\sigma = \sum_\nu \frac{\partial x'_\sigma}{\partial x_\nu} dx_\nu.$$

"(5)."

The dx'_σ's are expressed as linear and homogeneous function of dx_ν's; we can look upon the differentials of the co-ordinates as the components of a tensor, which we designate specially as a contravariant Four-vector.

Everything which is defined by Four quantities A^σ, with reference to a co-ordinate system, and transforms according to the same law,

$$A^{\sigma} = \sum_{\nu} \frac{\partial x'_{\sigma}}{\partial x_{\nu}} A^{\nu}$$

"(5a)."

we may call a contra-variant Four-vector. From (5. a), it follows at once that the sums ($A^\sigma \pm B^\sigma$) are also components of a four-vector, when A^σ and B^σ are so; corresponding relations hold also for all systems afterwards introduced as "tensors" (Rule of addition and subtraction of Tensors).

Co-variant Four-vector.

We call four quantities A_ν as the components of a covariant four-vector, when for any choice of the contra-variant four vector B^ν (6) $\sum_\nu A_\nu B^\nu =$ *Invariant.* From this definition follows the law of transformation of the co-variant four-vectors. If we substitute in the right hand side of the equation

$\sum_\sigma A'_\sigma B^{\sigma'} = \sum_\nu A_\nu B^\nu$.

the expressions

$$\sum_{\sigma} \frac{\partial x_{\nu}}{\partial x_{\sigma'}} B^{\sigma'}$$

for B^ν following from the inversion of the equation (5a) we get

$$\sum_{\sigma} B^{\sigma'} \sum_{\nu} \cdot \frac{\partial x_{\nu}}{\partial x_{\sigma'}} A_{\nu} = \sum_{\sigma} B^{\sigma'} A'_{\sigma}$$

As in the above equation $B^{\sigma'}$ are independent of one another and perfectly arbitrary, it follows that the transformation law is:—

$$A'_{\sigma} = \sum \frac{\partial x_{\nu}}{\partial x_{\sigma'}} A_{\nu}$$

Remarks on the simplification of the mode of writing the expressions. A glance at the equations of this paragraph will show that the indices which appear twice within the sign of summation [for example ν in (5)] are those over which the summation is to be made and that only over the indices which appear twice. It is therefore possible, without loss of clearness, to leave off the summation sign; so that we introduce the rule: wherever the index in any term of an expression appears twice, it is to be summed over all of them except when it is not expressedly said to the contrary.

The difference between the co-variant and the contra-variant four-vector lies in the transformation laws [(7) and (5)]. Both the quantities are tensors according to the above general remarks; in it lies its significance. In accordance with Ricci and Levi-civita, the contravariants and co-variants are designated by the over and under indices.

§ 6. Tensors of the second and higher ranks.

Contravariant tensor:—If we now calculate all the 16 products $A^{\mu\nu}$ of the components A^{μ} B^{ν}, of two contravariant four-vectors

(8) $A^{\mu\nu} = A^{\mu}B^{\nu}$

$A^{\mu\nu}$, will according to (8) and (5 a) satisfy the following transformation law.

$$A^{\sigma\tau'} = \frac{\partial x'_\sigma}{\partial x_\mu} \frac{\partial x'_\tau}{\partial x_\nu} A^{\mu\nu}$$

"(9)."

We call a thing which, with reference to any reference system is defined by 16 quantities and fulfils the transformation relation (9), a contravariant tensor of the second rank. Not every such tensor can be built from two four-vectors, (according to 8). But it is easy to show that any 16 quantities $A^{\mu\nu}$, can be represented as the sum of $A^\mu B^\nu$ of properly chosen four pairs of four-vectors. From it, we can prove in the simplest way all laws which hold true for the tensor of the second rank defined through (9), by proving it only for the special tensor of the type (8).

Contravariant Tensor of any rank:—It is clear that corresponding to (8) and (9), we can define contravariant tensors of the 3rd and higher ranks, with 4^3, etc. components. Thus it is clear from (8) and (9) that in this sense, we can look upon contravariant four-vectors, as contravariant tensors of the first rank.

Co-variant tensor.

If on the other hand, we take the 16 products $A_{\mu\nu}$ of the components of two co-variant four-vectors A_μ and B_ν,

(10) $A_{\mu\nu} = A_\mu B_\nu$.

for them holds the transformation law

$$A_{\sigma\tau'} = \frac{\partial x_\mu}{\partial x_{\sigma'}} \cdot \frac{\partial x'_\nu}{\partial x_{\tau'}} A_{\mu\nu}.$$

"(11)."

By means of these transformation laws, the co-variant tensor of the second rank is defined. All re-marks which we have already made concerning the contravariant tensors, hold also for co-variant tensors.

Remark:—

It is convenient to treat the scalar Invariant either as a contravariant or a co-variant tensor of zero rank.

Mixed tensor. We can also define a tensor of the second rank of the type

(12) $A_\mu^\nu = A_\mu B^\nu$

which is co-variant with reference to μ and contravariant with reference to ν. Its transformation law is

$$A_\sigma^{\tau'} = \frac{\partial x_{\tau'}}{\partial x_\beta} \cdot \frac{\partial x_\alpha}{\partial x_{\sigma'}} A_\alpha^\beta .$$

"(13)."

Naturally there are mixed tensors with any number of co-variant indices, and with any number of contra-variant indices. The co-variant and contra-variant tensors can be looked upon as special cases of mixed tensors.

Symmetrical tensors:—

A contravariant or a co-variant tensor of the second or higher rank is called symmetrical when any two components obtained by the mutual interchange of two indices are equal. The tensor $A^{\mu\nu}$ or $A_{\mu\nu}$ is symmetrical, when we have for any combination of indices

(14) $A^{\mu\nu} = A^{\nu\mu}$

or

(14a) $A_{\mu\nu} = A_{\nu\mu}$.

It must be proved that a symmetry so defined is a property independent of the system of reference. It follows in fact from (9) remembering (14)

$$A^{\sigma\tau'} = \frac{\partial x'_\sigma}{\partial x_\mu} \frac{\partial x'_\tau}{\partial x_\nu} A^{\mu\nu} = \frac{\partial x'_\sigma}{\partial x_\mu} \frac{\partial x'_\tau}{\partial x_\nu} A^{\nu\mu} = A^{\tau\sigma'}$$

Antisymmetrical tensor.

A contravariant or co-variant tensor of the 2nd, 3rd or 4th rank is called *antisymmetrical* when the two components got by mutually interchanging any two indices are equal and opposite. The tensor or $A^{\mu\nu}$ or $A_{\mu\nu}$ is thus antisymmetrical when we have

(15) $A^{\mu\nu} = -A^{\nu\mu}$

or

(15a) $A_{\mu\nu} = -A_{\nu\mu}$.

Of the 16 components $A^{\mu\nu}$, the four components $A^{\mu\mu}$ vanish, the rest are equal and opposite in pairs; so that there are only 6 numerically different components present (Six-vector).

Thus we also see that the antisymmetrical tensor $A^{\mu\nu\sigma}$ (3rd rank) has only 4 components numerically different, and the antisymmetrical tensor $A^{\mu\nu\sigma\tau}$ only one. Symmetrical tensors of ranks higher than the fourth, do not exist in a continuum of 4 dimensions.

§ 7. Multiplication of Tensors.

Outer multiplication of Tensors:—We get from the components of a tensor of rank z, and another of a rank z', the components of a tensor of rank $(z + z')$ for which we multiply all the components of the first with all the components of the second in pairs. For example, we obtain the tensor T from the tensors A and B of different kinds:—

$T_{\mu\nu\sigma} = A_{\mu\nu}B_\sigma$,

$T^{\alpha\beta\gamma\delta} = A^{\alpha\beta}B^{\gamma\delta}$,

$T_{\alpha\beta}{}^{\gamma\delta} = A_{\alpha\beta}B^{\gamma\delta}$.

The proof of the tensor character of T, follows immediately from the expressions (8), (10) or (12), or the transformation equations (9), (11), (13); equations (8), (10) and (12) are themselves examples of the outer multiplication of tensors of the first rank.

Reduction in rank of a mixed Tensor.

From every mixed tensor we can get a tensor which is two ranks lower, when we put an index of co-variant character equal to an index of the contravariant character and sum according to these indices (Reduction). We get for example, out of the mixed tensor of the fourth rank $A_{\alpha\beta}{}^{\gamma\delta}$, the mixed tensor of the second rank

$$A_{\beta}{}^{\delta} = A_{\alpha\beta}{}^{\alpha\delta} = (\textstyle\sum_{\alpha} A_{\alpha\beta}{}^{\alpha\delta})$$

and from it again by "reduction" the tensor of the zero rank

$$A = A_{\beta}{}^{\beta} = A_{\alpha\beta}{}^{\alpha\beta}.$$

The proof that the result of reduction retains a truly tensorial character, follows either from the representation of tensor according to the generalisation of (12) in combination with (6) or out of the generalisation of (13).

Inner and mixed multiplication of Tensors.

This consists in the combination of outer multiplication with reduction. Examples:—From the co-variant tensor of the second rank $A_{\mu\nu}$ and the contravariant tensor of the first rank B^{σ} we get by outer multiplication the mixed tensor

$$D^{\sigma}{}_{\mu\nu} = A_{\mu\nu} B^{\sigma} .$$

Through reduction according to indices ν and σ (*i.e.*, putting $\nu = \sigma$), the co-variant four vector

$D_{\mu} = D^{\nu}{}_{\mu\nu} = A_{\mu\nu} B^{\nu}$ is generated.

These we denote as the inner product of the tensor $A_{\mu\nu}$ and B^{σ}. Similarly we get from the tensors $A_{\mu\nu}$ and $B^{\sigma\tau}$ through outer multiplication and two-fold reduction the inner product $A_{\mu\nu} B^{\mu\nu}$. Through outer multiplication and one-fold reduction we get out of $A_{\mu\nu}$ and $B^{\sigma\tau}$, the mixed tensor of the second rank $D^{\tau}{}_{\mu} = A_{\mu\nu} B^{\tau\nu}$. We can fitly call this operation a mixed one; for it is outer with reference to the indices μ and τ and inner with respect to the indices ν and σ.

We now prove a law, which will be often applicable for proving the tensor-character of certain quantities. According to the above representation, $A_{\mu\nu} B^{\mu\nu}$ is a scalar, when $A_{\mu\nu}$ and $B^{\sigma\tau}$ are tensors. We also remark that when $A_{\mu\nu} B^{\mu\nu}$ is an invariant for every choice of the tensor $B^{\mu\nu}$, then $A_{\mu\nu}$ has a tensorial character.

Proof:—According to the above assumption, for any substitution we have

$A_{\sigma\tau'} B^{\sigma\tau'} = A_{\mu\nu} B^{\mu\nu}$.

From the inversion of (9) we have however

$$B^{\mu\nu} = \frac{\partial x_\mu}{\partial x_{\sigma'}} \frac{\partial x_\nu}{\partial x_{\tau'}} B^{\sigma\tau'}.$$

Substitution of this for $B^{\mu\nu}$ in the above equation gives

$$\left(A_{\sigma\tau'} - \frac{\partial x_\mu}{\partial x_{\sigma'}} \frac{\partial x_\nu}{\partial x_{\tau'}} A_{\mu\nu} \right) B^{\sigma\tau'} = 0.$$

This can be true, for any choice of $B^{\sigma\tau'}$ only when the term within the bracket vanishes. From which by referring to (11), the theorem at once follows. This law correspondingly holds for tensors of any rank and character. The proof is quite similar. The law can also be put in the following form. If B^μ and C^ν are any two vectors, and if for every choice of them the inner product $A_{\mu\nu} B^\mu C^\nu$ is a scalar, then $A_{\mu\nu}$ is a co-variant tensor. The last law holds even when there is the more special formulation, that with any arbitrary choice of the four-vector B^μ alone the scalar product $A_{\mu\nu}$ B^μ B^ν is a scalar, in which case we have the additional condition that $A_{\mu\nu}$ satisfies the symmetry condition. According to the method given above, we prove the tensor character of $(A_{\mu\nu} + A_{\nu\mu})$, from which on account of symmetry follows the tensor-character of $A_{\mu\nu}$. This law can easily be generalized in the case of co-variant and contravariant tensors of any rank.

Finally, from what has been proved, we can deduce the following law which can be easily generalized for any kind of tensor: If the quantities $A_{\mu\nu}$ B^ν form a tensor of the first rank, when B^ν is any arbitrarily chosen four-vector, then $A_{\mu\nu}$ is a tensor of the second rank. If for example, C^μ is any four-vector, then owing to the tensor character of $A_{\mu\nu}$ B^ν, the inner product $A_{\mu\nu}$ C^μ B^ν is a scalar, both the four-vectors C^μ and B^ν being arbitrarily chosen. Hence the proposition follows at once.

A few words about the Fundamental Tensor $g_{\mu\nu}$.

The co-variant fundamental tensor—In the invariant expression of the square of the linear element

$$ds^2 = g_{\mu\nu}\, dx_\mu\, dx_\nu$$

dx_μ plays the rôle of any arbitrarily chosen contravariant vector, since further $g_{\mu\nu} = g_{\nu\mu}$, it follows from the considerations of the last paragraph that $g_{\mu\nu}$ is a symmetrical co-variant tensor of the second rank. We call it the "fundamental tensor." Afterwards we shall deduce some properties of this tensor, which will also be true for any tensor of the second rank. But the special rôle of the fundamental tensor in our Theory, which has its physical basis on the particularly exceptional character of gravitation makes it clear that those relations are to be developed which will be required only in the case of the fundamental tensor.

The co-variant fundamental tensor.

If we form from the determinant scheme $|\ g_{\mu\nu}\ |$ the minors of $g_{\mu\nu}$ and divide them by the determinant $g = |\ g_{\mu\nu}\ |$ we get certain quantities $g^{\mu\nu} = g^{\nu\mu}$, which as we shall prove generates a contravariant tensor.

According to the well-known law of Determinants

$$(16)\ g_{\mu\sigma}\, g^{\nu\sigma} = \delta_\mu{}^\nu$$

where $\delta_\mu{}^\nu$ is 1, or 0, according as $\mu = \nu$ or not. Instead of the above expression for ds^2, we can also write

$$g_{\mu\sigma}\, \delta_\nu{}^\sigma\, dx_\mu\, dx_\nu$$

or according to (16) also in the form

$$g_{\mu\sigma} g_{\nu\tau} g^{\sigma\tau} dx_\mu dx_\nu$$

Now according to the rules of multiplication, of the fore-going paragraph, the magnitudes

$$d\xi_\sigma = g_{\mu\sigma} dx_\mu$$

forms a co-variant four-vector, and in fact (on account of the arbitrary choice of dx_μ) any arbitrary four-vector.

If we introduce it in our expression, we get

$$ds^2 = g^{\sigma\tau} d\xi_\sigma d\xi_\tau.$$

For any choice of the vectors $d\xi_\sigma$ $d\xi_\tau$ this is scalar, and $g^{\sigma\tau}$, according to its definition is a symmetrical thing in σ and τ, so it follows from the above results, that $g^{\sigma\tau}$ is a contravariant tensor. Out of (16) it also follows that δ^ν_μ is a tensor which we may call the mixed fundamental tensor.

Determinant of the fundamental tensor.

According to the law of multiplication of determinants, we have

$$| g_{\mu\alpha} g^{\alpha\nu} | = | g_{\mu\alpha} | \, | g^{\alpha\nu} |$$

On the other hand we have

$$| g_{\mu\alpha} g^{\alpha\nu} | = | \delta^\nu_\mu | = 1$$

So that it follows (17) that $| g_{\mu\nu} | \, | g^{\mu\nu} | = 1$.

Invariant of volume.

We see first the transformation law for the determinant $g = | g_{\mu\nu} |$. According to (11)

$$g' = \left| \frac{\partial x_{\mu}}{\partial x_{\sigma'}} \frac{\partial x_{\nu}}{\partial x_{\tau'}} g_{\mu\nu} \right|$$

From this by applying the law of multiplication twice, we obtain

$$g' = \left| \frac{\partial x_{\mu}}{\partial x_{\sigma'}} \right| \left| \frac{\partial x_{\nu}}{\partial x_{\tau'}} \right| \left| g_{\mu\nu} \right| = \left| \frac{\partial x_{\mu}}{\partial x_{\sigma'}} \right|^{2} g.$$

or

$$\sqrt{g'} = \left| \frac{\partial x_{\mu}}{\partial x_{\sigma'}} \right| \sqrt{g}$$

"(A)."

On the other hand the law of transformation of the volume element

$d\tau' = \int dx_1\, dx_2\, dx_3\, dx_4$

is according to the wellknown law of Jacobi.

$$d\tau' = \left| \frac{d\ '_{\sigma}}{dx_{\mu}} \right| d\tau.$$

"(B)."

by multiplication of the two last equations (A) and (B) we get

(18) $= \sqrt{g}\, d\tau' = \sqrt{g}\, d\tau$.

Instead of $\sqrt{g}$, we shall afterwards introduce $\sqrt{(-g)}$ which has a real value on account of the hyperbolic character of the time-space continuum. The invariant $\sqrt{(-g)}d\tau$, is equal in magnitude to the four-dimensional volume-element measured with solid rods and clocks, in accordance with the special relativity theory.

Remarks on the character of the space-time continuum—Our assumption that in an infinitely small region the special relativity theory holds, leads us to conclude that ds^2 can always, according to (1) be expressed in real magnitudes $dX_1 \ldots dX_b$. If we call $d\tau_0$ the "*natural*" volume element $dX_1\ dX_2\ dX_3\ dX_4$ we have thus (18a) $d\tau_0 = \sqrt{(g)}i\tau$.

Should $\sqrt{(-g)}$ vanish at any point of the four-dimensional continuum it would signify that to a finite co-ordinate volume at the place corresponds an infinitely small "natural volume." This can never be the case; so that g can never change its sign; we would, according to the special relativity theory assume that g has a finite negative value. It is a hypothesis about the physical nature of the continuum considered, and also a pre-established rule for the choice of co-ordinates.

If however $(-g)$ remains positive and finite, it is clear that the choice of co-ordinates can be so made that this quantity becomes equal to one. We would afterwards see that such a limitation of the choice of co-ordinates would produce a significant simplification in expressions for laws of nature.

In place of (18) it follows then simply that

$d\tau' = d$

from this it follows, remembering the law of Jacobi,

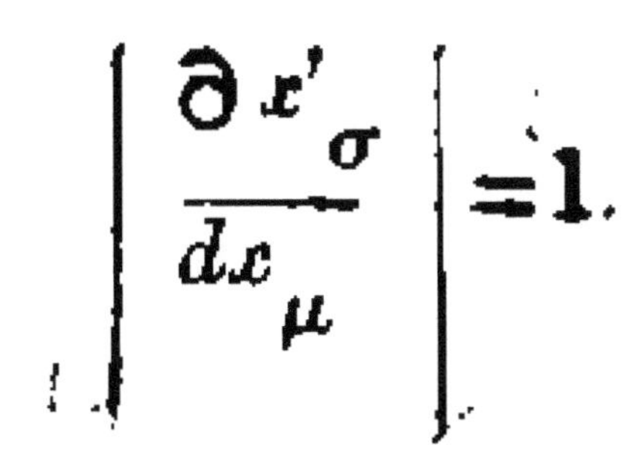

"(19)."

With this choice of co-ordinates, only substitutions with determinant 1 are allowable.

It would however be erroneous to think that this step signifies a partial renunciation of the general relativity postulate. We do not seek those laws of nature which are co-variants with regard to the transformations having the determinant 1, but we ask: what are the general co-variant laws of nature? First we get the law, and then we simplify its expression by a special choice of the system of reference.

Building up of new tensors with the help of the fundamental tensor.

Through inner, outer and mixed multiplications of a tensor with the fundamental tensor, tensors of other kinds and of other ranks can be formed.

Example:—

$A^{\mu} = g^{\mu\sigma} A_{\sigma}$

$A = g_{\mu\nu} A^{\mu\nu}$

We would point out specially the following combinations:

$A^{\mu\nu} = g^{\mu\alpha} g^{\nu\beta} A_{\alpha\beta}$

$$A_{\mu\nu} = g_{\mu\alpha}\, g_{\nu\beta}\, A^{\alpha\beta}$$

(complement to the co-variant or contravariant tensors)

and $B_{\mu\nu} = g_{\mu\nu}\, g^{\alpha\beta}\, A_{\alpha\beta}$

We can call $B_{\mu\nu}$ the reduced tensor related to $A_{\mu\nu}$.

Similarly

$$B^{\mu\nu} = g^{\mu\nu} g_{\alpha\beta} A^{\alpha\beta}.$$

It is to be remarked that $g^{\mu\nu}$ is no other than the "complement" of $g_{\mu\nu}$ for we have,—

$$g^{\mu\alpha} g^{\nu\beta} g_{\alpha\beta} = g_{\mu\alpha} \delta^{\nu}{}_{\alpha} = g^{\mu\nu}.$$

§ 9. Equation of the geodetic line (or of point-motion).

As the "line element" ds is a definite magnitude independent of the co-ordinate system, we have also between two points P_1 and P_2 of a four dimensional continuum a line for which $\int ds$ is an extremum (geodetic line), *i.e.*, one which has got a significance independent of the choice of co-ordinates.

Its equation is

$$(20)\quad \delta\left\{ \int_{P_1}^{P_2} ds \right\} = 0$$

From this equation, we can in a wellknown way deduce 4 total differential equations which define the geodetic line; this deduction is given here for the sake of completeness.

Let λ, be a function of the co-ordinates x_ν; this defines a series of surfaces which cut the geodetic line sought-for as well as all neighbouring lines from P_1 to P_2. We can suppose that all such curves are given when the value of its co-ordinates x_ν are given in terms of λ. The sign δ corresponds to a passage from a point of the geodetic curve sought-for to a point of the contiguous curve, both lying on the same surface λ.

Then (20) can be replaced by

$$(20a)\quad \int_{\lambda_1}^{\lambda_3} \delta\omega\, d\lambda = 0$$

{
{ $\omega^2 = g_{\mu\nu}(dx_\mu/d\lambda)(dx_\nu/d\lambda)$

But

$\delta\omega = (1/\omega)\{\tfrac{1}{2}(\partial g_{\mu\nu}/\partial x_\sigma) \cdot (dx_\mu/d\lambda) \cdot (dx_\nu/d\lambda) \cdot \delta x_\sigma$
$+ g_{\mu\nu}(dx_\mu/d\lambda)\delta(dx_\nu/d\lambda)\}$

So we get by the substitution of $\delta\omega$ in (20a), remembering that

$\delta(dx_\nu/d\lambda) = (d/d\lambda)(\delta x_\nu)$

after partial integration,

{ λ_3
{ $\int d\lambda\ k_\sigma\ \delta x_\sigma = 0$
(20b) { λ_1
{
{ where $k_\sigma = (d/d\lambda)\{(g_{\mu\nu}/\omega) \cdot (dx_\mu/d\lambda)\} - (1/(2\omega))(\partial g_{\mu\nu}/\partial x_\sigma$

$\times (dx_\mu/d\lambda) \cdot (dx_\nu/d\lambda)$.

From which it follows, since the choice of δv_σ is perfectly arbitrary that k_σ's should vanish. Then

(20c) $k_\sigma = 0$ ($\sigma = 1, 2, 3, 4$)

are the equations of geodetic line; since along the geodetic line considered we have $ds \neq 0$, we can choose the parameter λ, as the length of the arc measured along the geodetic line. Then $w = 1$, and we would get in place of (20c)

$$g_{\mu\nu}\frac{\partial^2 x_\mu}{\partial s^2} + \frac{\partial g_{\mu\nu}}{\partial x_\sigma}\frac{\partial x_\sigma}{\partial s}\frac{\partial x_\mu}{\partial s} - \frac{1}{2}\frac{\partial g_{\mu\sigma}}{\partial x_\nu}\frac{\partial x_\mu}{\partial s}\frac{\partial x_\sigma}{\partial s} = 0.$$

Or by merely changing the notation suitably,

$$g_{\alpha\sigma} \frac{d^2 x_\alpha}{ds^2} + \begin{bmatrix} \mu\nu \\ \sigma \end{bmatrix} \frac{dx_\mu}{ds} \cdot \frac{dx_\nu}{ds} = 0$$

"20d"

where we have put, following Christoffel,

$$\begin{bmatrix} \mu\nu \\ \sigma \end{bmatrix} = \frac{1}{2} \left[\frac{\partial g_{\mu\sigma}}{\partial x_\nu} + \frac{\partial g_{\nu\sigma}}{\partial x_\mu} - \frac{\partial g_{\mu\nu}}{\partial \sigma} \right].$$

"21"

Multiply finally (20d) with $g^{\sigma\tau}$ (outer multiplication with reference to τ, and inner with respect to σ) we get at last the final form of the equation of the geodetic line—

$$\frac{d^2 x_\tau}{ds^2} + \begin{Bmatrix} \mu\nu \\ \tau \end{Bmatrix} \frac{dx_\mu}{ds} \cdot \frac{dx_\nu}{ds} = 0.$$

Here we have put, following Christoffel,

$$\left\{ {\mu\nu \atop \tau} \right\} = g^{\tau\alpha} \left[{\mu\nu \atop \alpha} \right]$$

§ 10. Formation of Tensors through Differentiation.

Relying on the equation of the geodetic line, we can now easily deduce laws according to which new tensors can be formed from given tensors by differentiation. For this purpose, we would first establish the general co-variant differential equations. We achieve this through a repeated application of the following simple law. If a certain curve be given in our continuum whose points are characterised by the arc-distances *s*, measured from a fixed point on the curve, and if further φ, be an invariant space function, then *d*φ/*ds* is also an invariant. The proof follows from the fact that *d*φ as well as *ds*, are both invariants

Since

$$\frac{d\phi}{ds} = \frac{\partial \phi}{\partial x_\mu} \frac{\partial x_\mu}{\partial s}$$

so that

$$\psi = \frac{\partial \phi}{\partial x_\mu} \cdot \frac{dx_\mu}{ds}$$

is also an invariant for all curves which go out from a point in the continuum, *i.e.*, for any choice of the vector dx_μ. From which follows immediately that

$$A_\mu = \partial\varphi/\partial x_\mu$$

is a co-variant four-vector (gradient of φ).

According to our law, the differential-quotient $\chi = \partial\psi/\partial s$ taken along any curve is likewise an invariant.

Substituting the value of ψ, we get

$$\chi = \frac{\partial^2\phi}{\partial x_\mu \partial x_\nu} \cdot \frac{dx_\mu}{ds} \cdot \frac{dx_\nu}{ds} + \frac{\partial\phi}{\partial x_\mu} \cdot \frac{d^2 x_\mu}{ds^2}$$

Here however we can not at once deduce the existence of any tensor. If we however take that the curves along which we are differentiating are geodesics, we get from it by replacing d^2x_ν/ds^2 according to (22)

$$\chi = \left[\frac{\partial^2\phi}{\partial x_\mu \partial x_\nu} - \left\{ \begin{matrix} \mu\nu \\ \tau \end{matrix} \right\} \frac{\partial\phi}{\partial x_\tau} \right] \frac{dx_\mu}{ds} \cdot \frac{dx_\nu}{ds} \cdot$$

From the interchangeability of the differentiation with regard to μ and ν, and also according to (23) and (21) we see that the bracket

$$\left\{ \begin{matrix} \mu\nu \\ \tau \end{matrix} \right\}$$

is symmetrical with respect to μ and ν.

As we can draw a geodetic line in any direction from any point in the continuum, $\partial x_\mu / ds$ is thus a four-vector, with an arbitrary ratio of components, so that it follows from the results of §7 that

$$A_{\mu\nu} = \frac{\partial^2 \phi}{\partial x_\mu \partial x_\nu} - \left\{ \begin{matrix} \mu\nu \\ \tau \end{matrix} \right\} \frac{\partial \phi}{\partial x_\tau}$$

"25"

is a co-variant tensor of the second rank. We have thus got the result that out of the co-variant tensor of the first rank $A_\mu = \partial\varphi / \partial x_\mu$ we can get by differentiation a co-variant tensor of 2nd rank

$$A_{\mu\nu} = \frac{\partial A_\mu}{\partial x_\nu} - \left\{ \begin{matrix} \mu\nu \\ \tau \end{matrix} \right\} A_\tau$$

"26"

We call the tensor $A_{\mu\nu}$ the "extension" of the tensor A_μ. Then we can easily show that this combination also leads to a tensor, when the vector A_μ is not representable as a gradient. In order to see this we first remark that ψ $(d\varphi / \partial x_\mu)$ is a co-variant four-vector when ψ and φ are scalars. This is also the case for a sum of four such terms :—

$$S_\mu = \psi^{(1)} \frac{\partial \phi^{(1)}}{\partial x_\mu} + \dots + \psi^{(4)} \frac{\partial \phi^{(4)}}{\partial x_\mu}$$

when $\psi^{(1)}$, $\varphi^{(1)}$ ... $\psi^{(4)}$, $\varphi^{(4)}$ are scalars. Now it is however clear that every co-variant four-vector is representable in the form of S_μ.

If for example, A_μ is a four-vector whose components are any given functions of x_ν, we have, (with reference to the chosen co-ordinate system) only to put

$\psi^{(1)} = A_1 \; \varphi^{(1)} = x_1$

$\psi^{(2)} = A_2 \; \varphi^{(2)} = x_2$

$\psi^{(3)} = A_3 \; \varphi^{(3)} = x_3$

$\psi^{(4)} = A_4 \; \varphi^{(4)} = x_4$.

in order to arrive at the result that S_μ is equal to A_μ.

In order to prove then that $A_{\mu\nu}$ is a tensor when on the right side of (26) we substitute any co-variant four-vector for A_μ we have only to show that this is true for the four-vector S_μ. For this latter case, however, a glance on the right hand side of (26) will show that we have only to bring forth the proof for the case when

$A_\mu = \psi \, \partial\varphi / \partial x_\mu$.

Now the right hand side of (25) multiplied by ψ is

$$\psi \frac{\partial^2 \phi}{\partial x_\mu \partial x_\nu} - \begin{Bmatrix} \mu\nu \\ \tau \end{Bmatrix} \psi \frac{\partial \phi}{\partial x_\tau}$$

which has a tensor character. Similarly, $(\partial\varphi/\partial x_\mu)$ $(\partial\varphi/\partial x_\nu)$ is also a tensor (outer product of two four-vectors).

Through addition follows the tensor character of

$$\frac{\partial}{\partial x_\nu}\left(\psi\frac{\partial\phi}{\partial x_\mu}\right)-\left\{\begin{matrix}\mu\nu\\ \tau\end{matrix}\right\}\left(\psi\frac{\partial\phi}{\partial x_\tau}\right)$$

Thus we get the desired proof for the four-vector, $\psi\,\partial\varphi/\partial x_\mu$ and hence for any four-vectors A_μ as shown above.

With the help of the extension of the four-vector, we can easily define "extension" of a co-variant tensor of any rank. This is a generalisation of the extension of the four-vector. We confine ourselves to the case of the extension of the tensors of the 2nd rank for which the law of formation can be clearly seen.

As already remarked every co-variant tensor of the 2nd rank can be represented as a sum of the tensors of the type A_μ B_ν.

It would therefore be sufficient to deduce the expression of extension, for one such special tensor. According to (26) we have the expressions

$$\frac{\partial A_\mu}{\partial x_\sigma} - \left\{ \begin{matrix} \sigma\mu \\ \tau \end{matrix} \right\} A_\tau$$

$$\frac{\partial B_\nu}{\partial x_\sigma} - \left\{ \begin{matrix} \sigma\nu \\ \tau \end{matrix} \right\} B_\tau$$

are tensors. Through outer multiplication of the first with B_ν and the 2nd with A_μ we get tensors of the third rank. Their addition gives the tensor of the third rank

$$A_{\mu\nu\sigma} = \frac{\partial A_{\mu\nu}}{\partial x_\sigma} - \left\{ \begin{matrix} \sigma\mu \\ \tau \end{matrix} \right\} A_{\tau\nu} - \left\{ \begin{matrix} \sigma\nu \\ \tau \end{matrix} \right\} A_{\mu\tau}$$

"(27)"

where $A_{\mu\nu}$ is put $= A_\mu B_\nu$. The right hand side of (27) is linear and homogeneous with reference to $A_{\mu\nu}$, and its first differential co-efficient, so that this law of formation leads to a tensor not only in the case of a tensor of the type $A_\mu B_\nu$ but also in the case of a summation for all such tensors, *i.e.*, in the case of any co-variant tensor of the second rank. We call $A_{\mu\nu\sigma}$ the extension of the tensor $A_{\mu\nu}$. It is clear that (26) and (24) are only special cases of (27) (extension of the tensors of the first and zero rank). In general we can get all special laws of formation of tensors from (27) combined with tensor multiplication.

Some special cases of Particular Importance.

A few auxiliary lemmas concerning the fundamental tensor. We shall first deduce some of the lemmas much used afterwards. According to the law of differentiation of determinants, we have

(28) $dg = g^{\mu\nu} g \, dg_{\mu\nu} = -g_{\mu\nu} g dg^{\mu\nu}$.

The last form follows from the first when we remember that

$g_{\mu\nu} g^{\mu'\nu} = \delta^{\mu'}_{\mu}$, and therefore $g_{\mu\nu} g^{\mu\nu} = 4$,

consequently $g_{\mu\nu} dg^{\mu\nu} + g^{\mu\nu} dg_{\mu\nu} = 0$.

From (28), it follows that

$$\frac{1}{\sqrt{-g}} \frac{\partial \sqrt{-g}}{\partial x_\sigma} = \frac{1}{2} \frac{\partial \log(-g)}{\partial x_\sigma} = \frac{1}{2} g^{\mu\nu} \frac{\partial g_{\mu\nu}}{\partial x_\sigma}$$

$$= -\frac{1}{2} g_{\mu\nu} \frac{\partial g^{\mu\nu}}{\partial x_\sigma}.$$

"(29)"

Again, since $g_{\mu\nu} g^{\nu\sigma} = \delta^{\nu}_{\mu}$, we have, by differentiation,

$$\begin{cases} g_{\mu\sigma}\, dg^{\nu\sigma} = -g^{\nu\sigma}\, dg_{\mu\sigma} \\ \text{or} \\ g_{\mu\sigma} \dfrac{\partial g^{\nu\sigma}}{\partial x_\lambda} = -g^{\nu\sigma} \dfrac{\partial g_{\mu\sigma}}{\partial x_\lambda} \end{cases}$$

By mixed multiplication with $g^{\sigma\tau}$ and $g_{\nu\lambda}$ respectively we obtain (changing the mode of writing the indices).

$dg^{\mu\nu} = -g^{\mu\alpha}\, g^{\nu\beta}\, dg_{\alpha\beta}$

$\partial g^{\mu\nu}/\partial x_\sigma = -g^{\mu\alpha}\, g^{\nu\beta}\, dg_{\alpha\beta}$

and

(32)

$dg_{\mu\nu} = -g_{\mu\alpha}\, g_{\nu\beta}\, dg^{\alpha\beta}$

$\partial g_{\mu\nu}/\partial x_\sigma = -g_{\mu\alpha}\, g_{\nu\beta}\, \partial g^{\alpha\beta}/\partial x_\sigma$.

The expression (31) allows a transformation which we shall often use; according to (21)

$$\frac{\partial g_{\alpha\beta}}{\partial x_\sigma} = \begin{bmatrix} \alpha & \sigma \\ & \beta \end{bmatrix} + \begin{bmatrix} \beta & \sigma \\ & \alpha \end{bmatrix}$$

"(33)"

If we substitute this in the second of the formula (31), we get, remembering (23),

$$\frac{\partial g^{\mu\nu}}{\partial x_\sigma} = -\left(g^{\mu\tau} \begin{Bmatrix} \tau & \sigma \\ & \nu \end{Bmatrix} + g^{\nu\tau} \begin{Bmatrix} \tau & \sigma \\ & \mu \end{Bmatrix} \right)$$

"(34)"

By substituting the right-hand side of (34) in (29), we get

$$\frac{1}{\sqrt{-g}} \frac{\partial \sqrt{-g}}{\partial x_\sigma} = \begin{Bmatrix} \mu & \sigma \\ & \mu \end{Bmatrix}.$$

"(29a)"

Divergence of the contravariant four-vector.

Let us multiply (26) with the contravariant fundamental tensor $g^{\mu\nu}$ (inner multiplication), then by a transformation of the first member, the right-hand side takes the form

$$\frac{\partial}{\partial x_\nu}\left(g^{\mu\nu} A_\mu \right) - A_\mu \frac{\partial g^{\mu\nu}}{\partial x_\nu} - \frac{1}{2} g^{\tau\alpha} \left(\frac{\partial g_{\mu\alpha}}{\partial x_\nu} + \frac{\partial g_{\nu\alpha}}{\partial x_\mu} - \frac{\partial g_{\mu\nu}}{\partial x_\alpha} \right) g^{\mu\nu} A_\tau$$

"(A)"

According to (31) and (29), the last member can take the form

$$\frac{1}{2}\frac{\partial g^{\tau\nu}}{\partial x_\nu} A_\tau + \frac{1}{2}\frac{\partial g^{\mu\tau}}{\partial x_\mu} A_\tau + \frac{1}{\sqrt{-g}}\frac{\partial \sqrt{-g}}{\partial x_\alpha} g^{\mu\alpha} A_\tau$$

"(B)"

Both the first members of the expression (B), and the second member of the expression (A) cancel each other, since the naming of the summation-indices is immaterial. The last member of (B) can then be united with first of (A). If we put

$g^{\mu\nu} A_\mu = A^\nu$,

where A^ν as well as A_μ are vectors which can be arbitrarily chosen, we obtain finally

$$\Phi = \frac{1}{\sqrt{-g}}\frac{\partial}{\partial x_\nu}\left(\sqrt{-g}\, A^\nu\right)$$

This scalar is the *Divergence* of the contravariant four-vector A^ν.

Rotation of the (covariant) four-vector.

The second member in (26) is symmetrical in the indices μ, and ν. Hence $A_{\mu\nu} - A_{\nu\mu}$ is an antisymmetrical tensor built up in a very simple manner. We obtain

$$(36)\quad B_{\mu\nu} = \frac{\partial A_\mu}{\partial x_\nu} - \frac{\partial A_\nu}{\partial x_\mu}$$

Antisymmetrical Extension of a Six-vector.

If we apply the operation (27) on an antisymmetrical tensor of the second rank $A_{\mu\{v^2\}}$ and form all the equations arising from the cyclic interchange of the indices μ, ν, σ, and add all them, we obtain a tensor of the third rank

(37) $B_{\mu\nu\sigma} = A_{\mu\nu\sigma} + A_{\nu\sigma\mu} + A_{\sigma\mu\nu}$

$$= \frac{\partial A_{\mu\nu}}{\partial x_\sigma} + \frac{\partial A_{\nu\sigma}}{\partial x_\mu} + \frac{\partial A_{\sigma\mu}}{\partial x_\nu}$$

from which it is easy to see that the tensor is antisymmetrical.

Divergence of the Six-vector.

If (27) is multiplied by $g^{\mu\alpha}\, g^{\nu\beta}$ (mixed multiplication), then a tensor is obtained. The first member of the right hand side of (27) can be written in the form

$$\frac{\partial}{\partial x_\sigma}\left(g^{\mu\alpha}\, g^{\nu\beta} A_{\mu\nu}\right) - g^{\mu\alpha}\frac{\partial g^{\nu\beta}}{\partial x_\sigma} A_{\mu\nu} - g^{\nu\beta}\frac{\partial g^{\mu\alpha}}{\partial x_\sigma} A_{\mu\nu}.$$

If we replace $g^{\mu\alpha}\, g^{\nu\beta} A_{\mu\nu\sigma}$ by $A_\sigma{}^{\alpha\beta}$, $g^{\mu\alpha}\, g^{\nu\beta} A_{\mu\nu}$ by $A^{\alpha\beta}$ and replace in the transformed first member

$\partial g^{\nu\beta}/\partial x_\sigma$ and $\partial g^{\mu\alpha}/\partial x_\sigma$

with the help of (34), then from the right-hand side of (27) there arises an expression with seven terms, of which four cancel. There remains

$$A_\sigma^{\alpha\beta} = \frac{\partial A^{\alpha\beta}}{\partial x_\sigma} + \begin{Bmatrix} \sigma & \kappa \\ & \alpha \end{Bmatrix} A^{\kappa\beta} + \begin{Bmatrix} \sigma & \kappa \\ & \beta \end{Bmatrix} A^{\alpha\kappa}.$$

"(38)"

This is the expression for the extension of a contravariant tensor of the second rank; extensions can also be formed for corresponding contravariant tensors of higher and lower ranks.

We remark that in the same way, we can also form the extension of a mixed tensor A_{μ}^{α}

$$A^{\alpha}_{\mu\sigma} = \frac{\partial A^{\alpha}_{\mu}}{\partial x_{\sigma}} - \left\{ \begin{matrix} \sigma & \mu \\ & \tau \end{matrix} \right\} A^{\alpha}_{\tau} + \left\{ \begin{matrix} \sigma & \tau \\ & \alpha \end{matrix} \right\} A^{\tau}_{\mu}$$

"(39)"

By the reduction of (38) with reference to the indices β and σ(inner multiplication with δ_{β}^{σ}), we get a contravariant four-vector

$$A^{\alpha} = \frac{\partial A^{\alpha\beta}}{\partial x_{\beta}} + \left\{ \begin{matrix} \beta & \kappa \\ & \beta \end{matrix} \right\} A^{\alpha\kappa} + \left\{ \begin{matrix} \beta & \kappa \\ & \alpha \end{matrix} \right\} A^{\kappa\beta}.$$

On the account of the symmetry of

$$\left\{ \begin{matrix} \beta & \kappa \\ & \alpha \end{matrix} \right\}$$

with reference to the indices β and ϰ, the third member of the right hand side vanishes when $A^{\alpha\beta}$ is an antisymmetrical tensor, which we assume here;

the second member can be transformed according to (29a); we therefore get

$$A^{\alpha} = \frac{1}{\sqrt{-g}} \frac{\partial(\sqrt{-g}\, A^{\alpha\beta})}{\partial x_{\beta}}$$

"(40)"

This is the expression of the divergence of a contravariant six-vector.

Divergence of the mixed tensor of the second rank.

Let us form the reduction of (39) with reference to the indices α and σ, we obtain remembering (29a)

$$\sqrt{-g}\, A_{\mu} = \frac{\partial\left(\sqrt{-g}\, A^{\sigma}_{\mu}\right)}{\partial x_{\sigma}} - \begin{Bmatrix} \sigma & \mu \\ & \tau \end{Bmatrix} \sqrt{-g}\, A^{\sigma}_{\tau}$$

"(41)"

If we introduce into the last term the contravariant tensor $A^{\rho\sigma} = g^{\rho\tau} A^{\sigma}_{\tau}$, it takes the form

$$-\begin{bmatrix} \sigma & \mu \\ & \rho \end{bmatrix} \sqrt{-g}\, A^{\rho\sigma}.$$

If further $A^{\varrho\sigma}$ or is symmetrical it is reduced to

$$-\frac{1}{2}\sqrt{-g}\;\frac{\partial g_{\rho\sigma}}{\partial x_{\mu}}A^{\rho\sigma}.$$

If instead of $A^{\varrho\sigma}$, we introduce in a similar way the symmetrical co-variant tensor $A_{\varrho\sigma} = g_{\varrho\alpha}\, g_{\sigma\beta}\, A^{\alpha\beta}$, then owing to (31) the last member can take the form

$$\frac{1}{2}\sqrt{-g}\;\frac{\partial g^{\rho\sigma}}{\partial x_{\mu}}A_{\rho\sigma}$$

In the symmetrical case treated, (41) can be replaced by either of the forms

$$\sqrt{-g}\ A_{\mu} = \frac{\partial \left(\sqrt{-g}\ A^{\sigma}_{\mu} \right)}{\partial x_{\sigma}} - \frac{1}{2} \frac{\partial g_{\rho\sigma}}{\partial x_{\mu}} \sqrt{-g}\ A^{\rho\sigma}$$

"(41a)"

or

$$\sqrt{-g}\ A_{\mu} = \frac{\partial \left(\sqrt{-g}\ A^{\sigma}_{\mu} \right)}{\partial x_{\sigma}} + \frac{1}{2} \frac{\partial g^{\rho\sigma}}{\partial x_{\mu}} \sqrt{-g}\ A_{\sigma\rho}$$

"(41b)"

which we shall have to make use of afterwards.

§12. The Riemann-Christoffel Tensor.

We now seek only those tensors, which can be obtained from the fundamental tensor $g^{\mu\nu}$ by differentiation alone. It is found easily. We put in (27) instead of any tensor $A^{\mu\nu}$ the fundamental tensor $g^{\mu\nu}$ and get from it a new tensor, namely the extension of the fundamental tensor. We can easily convince ourselves that this vanishes identically. We prove it in the following way; we substitute in (27)

$$A_{\mu\nu} = \frac{\partial A_\mu}{\partial x_\nu} - \begin{Bmatrix} \mu\nu \\ \rho \end{Bmatrix} A_\rho$$

i.e., the extension of a four-vector.

Thus we get (by slightly changing the indices) the tensor of the third rank

$$A_{\mu\sigma\tau} = \frac{\partial^2 A_\mu}{\partial x_\sigma \partial x_\tau} - \begin{Bmatrix} \mu\sigma \\ \rho \end{Bmatrix} \frac{\partial A_\rho}{\partial x_\tau} - \begin{Bmatrix} \mu\tau \\ \rho \end{Bmatrix} \frac{\partial A_\rho}{\partial x_\sigma} - \begin{Bmatrix} \sigma\tau \\ \rho \end{Bmatrix} \cdot \frac{\partial A_\mu}{\partial x_\rho}$$

$$+ \left[- \frac{\partial}{\partial x_\tau} \begin{Bmatrix} \mu\sigma \\ \rho \end{Bmatrix} + \begin{Bmatrix} \mu\tau \\ \alpha \end{Bmatrix} \begin{Bmatrix} \alpha\sigma \\ \rho \end{Bmatrix} + \begin{Bmatrix} \sigma\tau \\ \alpha \end{Bmatrix} \begin{Bmatrix} \alpha\mu \\ \rho \end{Bmatrix} \right] A_\rho.$$

We use these expressions for the formation of the tensor $A_{\mu\sigma\tau}$ - $A_{\mu\tau\sigma}$. Thereby the following terms in $A_{\mu\sigma\tau}$ cancel the corresponding terms in $A_{\mu\tau\sigma}$; the first member, the fourth member, as well as the member corresponding to the last term within the square bracket. These are all symmetrical in σ, and τ. The same is true for the sum of the second and third members. We thus get

$$A_{\mu\sigma\tau} - A_{\mu\tau\sigma} = B^\rho_{\mu\sigma\tau} A_\rho .$$

$$\left\{ \begin{aligned} B^\rho_{\mu\sigma\tau} = & - \frac{\partial}{\partial x_\tau} \begin{Bmatrix} \mu\sigma \\ \rho \end{Bmatrix} + \frac{\partial}{\partial x_\sigma} \begin{Bmatrix} \mu\tau \\ \rho \end{Bmatrix} \\ & - \begin{Bmatrix} \mu\sigma \\ \alpha \end{Bmatrix} \begin{Bmatrix} \alpha\tau \\ \rho \end{Bmatrix} + \begin{Bmatrix} \mu\tau \\ \alpha \end{Bmatrix} \begin{Bmatrix} \alpha\sigma \\ \rho \end{Bmatrix} \end{aligned} \right.$$

"(43)"

The essential thing in this result is that on the right hand side of (42) we have only A_ϱ, but not its differential co-efficients. From the tensor-character of $A_{\mu\sigma\tau} - A_{\mu\tau\sigma}$, and from the fact that A_ϱ is an arbitrary four vector, it follows, on account of the result of §7, that $B^\varrho_{\mu\sigma\tau}$ is a tensor (Riemann-Christoffel Tensor).

The mathematical significance of this tensor is as follows; when the continuum is so shaped, that there is a co-ordinate system for which $g_{\mu\nu}$'s are constants, $B^\varrho_{\mu\sigma\tau}$ all vanish.

If we choose instead of the original co-ordinate system any new one, so would the $g_{\mu\nu}$'s referred to this last system be no longer constants. The tensor character of $B^\varrho_{\mu\sigma\tau}$ shows us, however, that these components vanish collectively also in any other chosen system of reference. The vanishing of the Riemann Tensor is thus a necessary condition that for some choice of the axis-system $g_{\mu\nu}$'s can be taken as constants. In our problem it corresponds to the case when by a suitable choice of the co-ordinate system, the special relativity theory holds throughout any finite region. By the reduction of (43) with reference to indices to τ and ϱ, we get the covariant tensor of the second rank

$$B_{\mu\nu} = R_{\mu\nu} + S_{\mu\nu}$$

$$R_{\mu\nu} = -\frac{\partial}{\partial x_\alpha}\begin{Bmatrix}\mu\nu \\ \alpha\end{Bmatrix} + \begin{Bmatrix}\mu\alpha \\ \beta\end{Bmatrix}\begin{Bmatrix}\nu\beta \\ \alpha\end{Bmatrix}$$

$$S_{\mu\nu} = \frac{\partial \log \sqrt{-g}}{\partial x_\mu \, \partial x_\nu} - \begin{Bmatrix}\mu\nu \\ \alpha\end{Bmatrix}\frac{\partial \log \sqrt{-g}}{\partial x_\alpha}.$$

"(44)"

Remarks upon the choice of co-ordinates.—It has already been remarked in §8, with reference to the equation (18a), that the co-ordinates can with advantage be so chosen that $\sqrt{(-g)} = 1$. A glance at the equations got in the last two paragraphs shows that, through such a choice, the law of formation of the tensors suffers a significant simplification. It is specially true for the tensor $B_{\mu\nu}$, which plays a fundamental rôle in the theory. By this simplification, $S_{\mu\nu}$ vanishes of itself so that tensor $B_{\mu\nu}$ reduces to $R_{\mu\nu}$.

I shall give in the following pages all relations in the simplified form, with the above-named specialisation of the co-ordinates. It is then very easy to go back to the general covariant equations, if it appears desirable in any special case.

C. THE THEORY OF THE GRAVITATION-FIELD

§13. Equation of motion of a material point in a gravitation-field. Expression for the field-components of gravitation.

A freely moving body not acted on by external forces moves, according to the special relativity theory, along a straight line and uniformly. This also holds for the generalised relativity theory for any part of the four-dimensional region, in which the co-ordinates K_0 can be, and are, so chosen that $g_{\mu\nu}$'s have special constant values of the expression (4).

Let us discuss this motion from the stand-point of any arbitrary co-ordinate-system K_1; it moves with reference to K_1 (as explained in §2) in a gravitational field. The laws of motion with reference to K_1 follow easily from the following consideration. With reference to K_0, the law of motion is a four-dimensional straight line and thus a geodesic. As a geodetic-line is defined independently of the system of co-ordinates, it would also be the law of motion for the motion of the material-point with reference to K_1. If we put

$$\Gamma^{\tau}_{\mu\nu} = -\begin{Bmatrix} \mu\nu \\ \tau \end{Bmatrix}$$

"(45)"

we get the motion of the point with reference to K_1, given by

$$\frac{d^2 x_\tau}{ds^2} = \Gamma^{\tau}_{\mu\nu} \frac{dx_\mu}{ds} \frac{dx_\nu}{ds}.$$

"(46)"

We now make the very simple assumption that this general covariant system of equations defines also the motion of the point in the gravitational field, when there exists no reference-system K_0, with reference to which the special relativity theory holds throughout a finite region. The assumption seems to us to be all the more legitimate, as (46) contains only the first differentials of $g_{\mu\nu}$, among which there is no relation in the special case when K_0 exists.

If $\gamma_{\mu\nu}^{\tau}$'s vanish, the point moves uniformly and in a straight line; these magnitudes therefore determine the deviation from uniformity. They are the components of the gravitational field.

§14. The Field-equation of Gravitation in the absence of matter.

In the following, we differentiate gravitation-field from matter in the sense that everything besides the gravitation-field will be signified as matter; therefore the term includes not only matter in the usual sense, but also the electro-dynamic field. Our next problem is to seek the field-equations of gravitation in the absence of matter. For this we apply the same method as employed in the foregoing paragraph for the deduction of the equations of motion for material points. A special case in which the field-equations sought-for are evidently satisfied is that of the special relativity theory in which $g_{\mu\nu}$'s have certain constant values. This would be the case in a certain finite region with reference to a definite co-ordinate system K_0. With reference to this system, all the components $B^{\varrho}_{\mu\sigma\tau}$ of the Riemann's Tensor [equation 43] vanish. These vanish then also in the region considered, with reference to every other co-ordinate system.

The equations of the gravitation-field free from matter must thus be in every case satisfied when all $B^{\varrho}_{\mu\sigma\tau}$ vanish. But this condition is clearly one which goes too far. For it is clear that the gravitation-field generated by a material point in its own neighbourhood can never be transformed *away* by any choice of axes, *i.e.*, it cannot be transformed to a case of constant $g_{\mu\nu}$'s.

Therefore it is clear that, for a gravitational field free from matter, it is desirable that the symmetrical tensors $B_{\mu\nu}$ deduced from the tensors $B^{\varrho}_{\mu\sigma\tau}$ should vanish. We thus get 10 equations for 10 quantities $g_{\mu\nu}$ which are fulfilled in the special case when $B^{\varrho}_{\mu\sigma\tau}$'s all vanish.

Remembering (44) we see that in absence of matter the field-equations come out as follows; (when referred to the special co-ordinate-system chosen.)

$$\frac{\partial \Gamma^{\alpha}_{\mu\nu}}{\partial x_{\alpha}} + \Gamma^{\alpha}_{\mu\beta} \Gamma^{\beta}_{\nu\alpha} = 0;$$

$$\sqrt{-g} = 1; \quad \Gamma^{\alpha}_{\mu\nu} = -\left\{ \begin{matrix} \mu\nu \\ \alpha \end{matrix} \right\}.$$

"(47)"

It can also be shown that the choice of these equations is connected with a minimum of arbitrariness. For besides $B_{\mu\nu}$, there is no tensor of the second rank, which can be built out of $g_{\mu\nu}$'s and their derivatives no higher than the second, and which is also linear in them.

It will be shown that the equations arising in a purely mathematical way out of the conditions of the general relativity, together with equations (46), give us the Newtonian law of attraction as a first approximation, and lead in the second approximation to the explanation of the perihelion-motion of mercury discovered by Leverrier (the residual effect which could not be accounted for by the consideration of all sorts of disturbing factors). My view is that these are convincing proofs of the physical correctness of my theory.

§15. Hamiltonian Function for the Gravitation-field. Laws of Impulse and Energy.

In order to show that the field equations correspond to the laws of impulse and energy, it is most convenient to write it in the following Hamiltonian form:—

(47a)

$\delta\int H d\tau = 0$

$H = g^{\mu\nu} \gamma^{\alpha}_{\mu\beta} \gamma^{\beta}_{\nu\alpha}$

$\sqrt{(-g)} = 1$

Here the variations vanish at the limits of the finite four-dimensional integration-space considered.

It is first necessary to show that the form (47a) is equivalent to equations (47). For this purpose, let us consider H as a function of $g^{\mu\nu}$ and $g^{\mu\nu}_{\sigma}$ (= $\partial g^{\mu\nu}/\partial x_{\sigma}$)

We have at first

$\delta H = \Gamma^{\alpha}_{\mu\beta}\, \Gamma^{\beta}_{\nu\alpha}\, \delta g^{\mu\nu} + 2g^{\mu\nu}\, \Gamma^{\alpha}_{\mu\beta}\, \delta\Gamma^{\beta}_{\nu\alpha}$

$= - \Gamma^{\alpha}_{\mu\beta}\, \Gamma^{\beta}_{\nu\alpha}\, \delta g^{\mu\nu} + 2\Gamma^{\alpha}_{\mu\beta}\, \delta(g^{\mu\nu}\Gamma^{\beta}_{\nu\alpha})$.

But

$$\delta\left(g^{\mu\nu}\Gamma^{\beta}_{\nu\alpha}\right) = -\frac{1}{2}\, \delta\left[g^{\mu\nu}\, g^{\beta\lambda}\left(\frac{\partial g_{\nu\lambda}}{\partial x_{\alpha}} + \frac{\partial g_{\alpha\lambda}}{\partial x_{\nu}} - \frac{\partial g_{\alpha\nu}}{\partial x_{\lambda}}\right)\right.$$

The terms arising out of the two last terms within the round bracket are of different signs, and change into one another by the interchange of the indices μ and β. They cancel each other in the expression for δH, when they are multiplied by $\Gamma_{\mu\beta}^{\alpha}$, which is symmetrical with respect to μ and β, so that only the first member of the bracket remains for our consideration. Remembering (31), we thus have:—

$\delta H = -\Gamma_{\mu\beta}^{\alpha}\, \Gamma_{\nu\alpha}^{\beta}\, \delta g^{\mu\nu} + \Gamma_{\mu\beta}^{\alpha}\, \delta g_{\alpha}^{\mu\beta}$

Therefore

(48)

$\partial H/\partial g^{\mu\nu} = -\Gamma_{\mu\beta}^{\alpha}\, \Gamma_{\nu\alpha}^{\beta}$

$\partial H/\partial g_{\sigma}^{\mu\nu} = \Gamma_{\mu\nu}^{\sigma}$

If we now carry out the variations in (47a), we obtain the system of equations

(47b) $\partial/\partial x_{\alpha}\, (\, \partial H/\partial g_{\alpha}^{\mu\nu}\,) - \partial H/\partial g^{\mu\nu} = 0$, which, owing to the relations (48), coincide with (47), as was required to be proved.

If (47b) is multiplied by $g_\sigma^{\mu\nu}$, since

$$\partial g_\sigma^{\mu\nu}/\partial x_\alpha = \partial g_\alpha^{\mu\nu}/\partial x_\sigma$$

and consequently

$$g_\sigma^{\mu\nu}\, \partial/\partial x_\alpha\, (\partial H/\partial g_\alpha^{\mu\nu}) = \partial/\partial x_\alpha\, (g_\sigma^{\mu\nu}\, \partial H/\partial g_\alpha^{\mu\nu})$$
$$- \partial H/\partial g_\alpha^{\mu\nu}\, \partial g_\alpha^{\mu\nu}/\partial x_\sigma$$

we obtain the equation

$$\partial/\partial x_\alpha\, (g_\sigma^{\mu\nu}\, \partial H/\partial g_\alpha^{\mu\nu}) - \partial H/\partial x_\sigma = 0$$

or

$$\{\ \partial t_\sigma^\alpha/\partial x_\alpha = 0$$

$$(49)\ \{\ -2\varkappa t_\sigma^\alpha = g_\sigma^{\mu\nu}\, \partial H/\partial g_\alpha^{\mu\nu} - \delta_\sigma^\alpha\, H.$$

Owing to the relations (48), the equations (47) and (34),

$$(50)\ \varkappa t_\sigma^\alpha = \tfrac{1}{2}\, \delta_\sigma^\alpha\, g^{\mu\nu}\, \Gamma_{\mu\beta}^\alpha\, \Gamma_{\nu\alpha}^\beta$$
$$- g^{\mu\nu}\, \Gamma_{\mu\beta}^\alpha\, \Gamma_{\nu\sigma}^\beta.$$

It is to be noticed that t_σ^α is not a tensor, so that the equation (49) holds only for systems for which $\sqrt{-g} = 1$. This equation expresses the laws of conservation of impulse and energy in a gravitation-field. In fact, the integration of this equation over a three-dimensional volume V leads to the four equations

$$(49a)\ d/dx_4\, \{\textstyle\int t_\sigma^4\, dV\} = \int (t_\sigma^1\, \alpha_1$$
$$+ t_\sigma^2\, \alpha_2 + t_\sigma^3\, \alpha_3) dS$$

where α_1, α_2, α_2 are the direction-cosines of the inward-drawn normal to the surface-element dS in the Euclidean Sense. We recognise in this the usual expression for the laws of conservation. We denote the magnitudes $t^\alpha{}_\sigma$ as the energy-components of the gravitation-field.

I will now put the equation (47) in a third form which will be very serviceable for a quick realisation of our object. By multiplying the field-equations (47) with $g^{\nu\sigma}$, these are obtained in the mixed forms. If we remember that

$$g^{\nu\sigma} \partial\Gamma^{\alpha}_{\mu\nu}/\partial x_{\alpha} = \partial/\partial x_{\alpha} (g^{\nu\sigma} \Gamma^{\alpha}_{\mu\nu}) - \partial g^{\nu\sigma}/\partial x_{\alpha} \Gamma^{\alpha}_{\mu\nu},$$

which owing to (34) is equal to

$$\partial/\partial x_{\alpha} (.g^{\nu\sigma} \Gamma^{\alpha}_{\mu\nu}) - g^{\nu\beta} \Gamma^{\sigma}_{\alpha\beta} \Gamma\gamma^{\alpha}_{\mu\nu}$$

$$- g^{\sigma\beta} \Gamma^{\nu}_{\beta\alpha} \Gamma^{\alpha}_{\mu\nu},$$

or slightly altering the notation, equal to

$$\partial/\partial x_{\alpha} (g^{\sigma\beta} \Gamma^{\alpha}_{\mu\beta}) - g^{mn} \Gamma^{\sigma}_{m\beta} \Gamma^{\beta}_{n\mu}$$

$$- g^{\nu\sigma} \Gamma^{\alpha}_{\mu\beta} \Gamma^{\beta}_{\nu\alpha}.$$

The third member of this expression cancels with the second member of the field-equations (47). In place of the second term of this expression, we can, on account of the relations (50), put

$\varkappa (t^{\sigma}_{\mu} - \frac{1}{2} \delta^{\sigma}_{\mu} t)$, where $t = t^{\alpha}_{\alpha}$

Therefore in the place of the equations (47), we obtain

$$(51) \{ \partial/\partial x_{\alpha} (g^{\sigma\beta} \Gamma^{\alpha}_{\mu\beta}) = -\varkappa(t^{\sigma}_{\mu} - \tfrac{1}{2} \delta^{\sigma}_{\mu} t)$$

$$\{ \sqrt{(-g)} = 1.$$

§16. General formulation of the field-equation of Gravitation.

The field-equations established in the preceding paragraph for spaces free from matter is to be compared with the equation $\nabla^2\varphi = 0$ of the Newtonian theory. We have now to find the equations which will correspond to Poisson's Equation $\nabla^2\varphi = 4\pi\varkappa\varrho$ (ϱ signifies the density of matter).

The special relativity theory has led to the conception that the inertial mass (Träge Masse) is no other than energy. It can also be fully expressed mathematically by a symmetrical tensor of the second rank, the energy-tensor. We have therefore to introduce in our generalised theory energy-tensor τ^{α}_{σ} associated with matter, which like the energy components t^{α}_{σ} of the gravitation-field (equations 49, and 50) have a mixed character but which however can be connected with symmetrical covariant tensors. The equation (51) teaches us how to introduce the energy-tensor (corresponding to the density of Poisson's equation) in the field equations of gravitation. If we consider a complete system (for example the Solar-system) its total mass, as also its total gravitating action, will depend on the total energy of the system, ponderable as well as gravitational. This can be expressed, by putting in (51), in place of energy-components t_μ^σ of gravitation-field alone the sum of the energy-components of matter and gravitation, *i.e.*,

$t_\mu^\sigma + T_\mu^\sigma$.

We thus get instead of (51), the tensor-equation

$$\frac{\partial}{\partial x_\alpha}\left(g^{\sigma\beta}\Gamma^\alpha_{\mu\beta}\right) = -\kappa\left[\left(t^\sigma_\mu + T^\sigma_\mu\right) - \frac{1}{2}\delta^\sigma_\mu (t + T)\right]$$

$$\sqrt{-g} = 1$$

"(52)"

where $T = T_\mu^\mu$ (Laue's Scalar). These are the general field-equations of gravitation in the mixed form. In place of (47), we get by working backwards the system

$$\frac{\partial \Gamma^\alpha_{\mu\nu}}{\partial x_\alpha} + \Gamma^\alpha_{\mu\beta}\Gamma^\beta_{\nu\alpha} = -\kappa\left(T_{\mu\nu} - \frac{1}{2}g_{\mu\nu}T\right)$$

$$\sqrt{-g} = 1.$$

"(53)"

It must be admitted, that this introduction of the energy-tensor of matter cannot be justified by means of the Relativity-Postulate alone; for we have in the foregoing analysis deduced it from the condition that the energy of the gravitation-field should exert gravitating action in the same way as every other kind of energy. The strongest ground for the choice of the above equation however lies in this, that they lead, as their consequences, to equations expressing the conservation of the components of total energy (the impulses and the energy) which exactly correspond to the equations (49) and (49a). This shall be shown afterwards.

§17. The laws of conservation in the general case.

The equations (52) can be easily so transformed that the second member on the right-hand side vanishes. We reduce (52) with reference to the indices μ and σ and subtract the equation so obtained after multiplication with $\frac{1}{2}\delta_\mu^\sigma$ from (52).

We obtain,

$$(52a)\ \partial/\partial x_\alpha(g^{\sigma\beta}\Gamma_{\mu\beta}^\alpha - \tfrac{1}{2}\delta_\mu^\sigma g^{\lambda\beta}\Gamma_{\lambda\beta}^\alpha)$$
$$= -\varkappa(t_\mu^\sigma + T_\mu^\sigma)$$

we operate on it by $\partial/\partial x_\sigma$. Now,

$$\partial^2/\partial x_\alpha \partial x_\sigma\ (g^{\sigma\beta}\Gamma_{\mu\beta}^\alpha)$$
$$= -\tfrac{1}{2}\,\partial^2/\partial x_\alpha \partial x_\sigma\ [g^{\sigma\beta} g^{\alpha\lambda}(\partial g_{\mu\lambda}/\partial x_\beta$$
$$+ \partial g_{\beta\lambda}/\partial x_\mu - \partial g_{\mu\beta}/\partial x_\lambda)].$$

The first and the third member of the round bracket lead to expressions which cancel one another, as can be easily seen by interchanging the summation-indices α, and σ, on the one hand, and β and λ, on the other.

The second term can be transformed according to (31). So that we get,

$$(54)\ \partial^2/\partial x_\alpha \partial x_\sigma\ (g^{\sigma\beta}\gamma_{\mu\beta}^\alpha)$$
$$= \tfrac{1}{2}\,\partial^3 g^{\alpha\beta}/\partial x_\sigma \partial x_\beta \partial x_\mu$$

The second member of the expression on the left-hand side of (52a) leads first to

$$-\tfrac{1}{2}\,\partial^2/\partial x_\alpha \partial x_\mu\ (g^{\lambda\beta}\Gamma_{\lambda\beta}^\alpha) \text{ or}$$

$$\text{to } 1/4\ \partial^2/\partial x_\alpha \partial x_\mu\ [g^{\lambda\beta}g^{\alpha\delta}(\ \partial g_{\delta\lambda}/\partial x_\beta$$
$$+ \partial g_{\delta\beta}/\partial x_\lambda - \partial g_{\lambda\beta}/\partial x_\delta)].$$

The expression arising out of the last member within the round bracket vanishes according to (29) on account of the choice of axes. The two others can be taken together and give us on account of (31), the expression

$-\frac{1}{2}\, \partial^3 g^{\alpha\beta} / \partial x_\alpha \partial x_\beta \partial x_\mu$

So that remembering (54) we have

(55) $\partial^2 / \partial x_\alpha \partial x_\sigma \, (g^{\sigma\beta} \Gamma_{\mu\beta}{}^{\alpha}$

$- \frac{1}{2}\, \delta_\mu{}^\sigma \, g^{\lambda\beta} \, \Gamma_{\lambda\beta}{}^{\alpha}) = 0.$

identically.

From (55) and (52a) it follows that

(56) $\partial / \partial x_\sigma \, (t_\mu{}^\sigma + T_\mu{}^\sigma) = 0$

From the field equations of gravitation, it also follows that the conservation-laws of impulse and energy are satisfied. We see it most simply following the same reasoning which lead to equations (49a); only instead of the energy-components of the gravitational-field, we are to introduce the total energy-components of matter and gravitational field.

§18. The Impulse-energy law for matter as a consequence of the field-equations.

If we multiply (53) with $\partial g^{\mu\nu} / \partial x_\sigma$, we get in a way similar to §15, remembering that

$g_{\mu\nu} \, \partial g^{\mu\nu} / \partial x_\sigma$ vanishes,

the equations $\partial t_\sigma{}^\alpha / \partial x_\alpha - \frac{1}{2}\, \partial g^{\mu\nu} / \partial x_\sigma \, T_{\mu\nu} = 0$

or remembering (56)

(57) $\partial T_\sigma{}^\alpha / \partial x_\alpha + \frac{1}{2}\, \partial g^{\mu\nu} / \partial x_\sigma \, T_{\mu\nu} = 0$

A comparison with (41b) shows that these equations for the above choice of co-ordinates ($\sqrt{(-g)} = 1$) asserts nothing but the vanishing of the divergence of the tensor of the energy-components of matter.

Physically the appearance of the second term on the left-hand side shows that for matter alone the law of conservation of impulse and energy cannot hold; or can only hold when $g^{\mu\nu}$'s are constants; *i.e.*, when the field of gravitation vanishes. The second member is an expression for impulse and energy which the gravitation-field exerts per time and per volume upon matter. This comes out clearer when instead of (57) we write it in the form of (47).

(57a) $\partial T_\sigma^\alpha / \partial x_\alpha = -\Gamma_{\sigma\beta}^\alpha T_\alpha^\beta$.

The right-hand side expresses the interaction of the energy of the gravitational-field on matter. The field-equations of gravitation contain thus at the same time 4 conditions which are to be satisfied by all material phenomena. We get the equations of the material phenomena completely when the latter is characterised by four other differential equations independent of one another.

D. THE "MATERIAL" PHENOMENA.

The Mathematical auxiliaries developed under 'B' at once enables us to generalise, according to the generalised theory of relativity, the physical laws of matter (Hydrodynamics, Maxwell's Electro-dynamics) as they lie already formulated according to the special-relativity-theory. The generalised Relativity Principle leads us to no further limitation of possibilities; but it enables us to know exactly the influence of gravitation on all processes without the introduction of any new hypothesis.

It is owing to this, that as regards the physical nature of matter (in a narrow sense) no definite necessary assumptions are to be introduced. The question may lie open whether the theories of the electro-magnetic field and the gravitational-field together, will form a sufficient basis for the theory of matter. The general relativity postulate can teach us no new principle. But by building up the theory it must be shown whether electro-magnetism and gravitation together can achieve what the former alone did not succeed in doing.

§19. Euler's equations for frictionless adiabatic liquid.

Let p and ϱ, be two scalars, of which the first denotes the pressure and the last the density of the fluid; between them there is a relation. Let the contravariant symmetrical tensor

$T^{\alpha\beta} = -g^{\alpha\beta} p + \varrho\, dx_\alpha/ds\, dx_\beta/ds$ (58)

be the contra-variant energy-tensor of the liquid. To it also belongs the covariant tensor

(58a) $T_{\mu\nu} = -g_{\mu\nu} p + g_{\mu\alpha}\, dx_\alpha/ds\, g_{\mu\beta}\, dx_\beta/ds\, \varrho$

as well as the mixed tensor

(58b) $T^\alpha_\sigma = -\delta^\alpha_\sigma p + g_{\sigma\beta}\, dx_\beta/ds\, dx_\alpha/ds\, \varrho$.

If we put the right-hand side of (58b) in (57a) we get the general hydrodynamical equations of Euler according to the generalised relativity theory. This in principle completely solves the problem of motion; for the four equations (57a) together with the given equation between p and ϱ, and the equation

$g_{\alpha\beta}\, dx\_\alpha/ds\, dx_\beta/ds = 1$,

are sufficient, with the given values of $g_{\alpha\beta}$, for finding out the six unknowns

$p, \varrho, dx_1/ds, dx_2/ds, dx_3/ds\ dx_4/ds.$

If $g_{\mu\nu}$'s are unknown we have also to take the equations (53). There are now 11 equations for finding out 10 functions g, so that the number is more than sufficient. Now it is be noticed that the equation (57a) is already contained in (53), so that the latter only represents (7) independent equations. This indefiniteness is due to the wide freedom in the choice of co-ordinates, so that mathematically the problem is indefinite in the sense that three of the space-functions can be arbitrarily chosen.

§20. Maxwell's Electro-Magnetic field-equations.

Let φ_ν be the components of a covariant four-vector, the electro-magnetic potential; from it let us form according to (36) the components $F_{\varrho\sigma}$ of the covariant six-vector of the electro-magnetic field according to the system of equations

(59) $F_{\varrho\sigma} = \partial\varphi_\varrho/\partial x_\sigma - \partial\varphi_\sigma/\partial x_\varrho$.

From (59), it follows that the system of equations

(60) $\partial F_{\varrho\sigma}/\partial x_\tau + \partial F_{\sigma\tau}/\partial x_\varrho + \partial F_{\tau\varrho}/\partial x_\sigma = 0$

is satisfied of which the left-hand side, according to (37), is an anti-symmetrical tensor of the third kind. This system (60) contains essentially four equations, which can be thus written:—

{ $\partial F_{23}/\partial x_4 + \partial F_{34}/\partial x_2\ \partial F_{42}/\partial x_3 = 0$

{

{ $\partial F_{34}/\partial x_1 + \partial F_{41}/\partial x_3\ \partial F_{13}/\partial x_4 = 0$

(60a) {

{ $\partial F_{41}/\partial x_2 + \partial F_{12}/\partial x_4\ \partial F_{24}/\partial x_1 = 0$

{

{ $\partial F_{12}/\partial x_3 + \partial F_{23}/\partial x_1\ \partial F_{31}/\partial x_2 = 0$.

This system of equations corresponds to the second system of equations of Maxwell. We see it at once if we put

{ $F_{23} = H_x\ F_{14} = E_x$

{

(61) $\{ F_{31} = H_y \; F_{24} = E_y$

$\{$

$\{ F_{12} = H_z \; F_{34} = E_z$

Instead of (60a) we can therefore write according to the usual notation of three-dimensional vector-analysis:—

$\{ \partial H/\partial t + \text{rot } E = 0$

(60b) $\{$

$\{ \text{div } H = 0.$

The first Maxwellian system is obtained by a generalisation of the form given by Minkowski.

We introduce the contra-variant six-vector $F_{\alpha\beta}$ by the equation

(62) $F^{\mu\nu} = g^{\mu\alpha} g^{\nu\beta} F_{\alpha\beta}$,

and also a contra-variant four-vector J^{μ}, which is the electrical current-density in vacuum. Then remembering (40) we can establish the system of equations, which remains invariant for any substitution with determinant 1 (according to our choice of co-ordinates).

(63) $\partial F^{\mu\nu}/\partial x_{\nu} = J^{\mu}$

If we put

$\{ F^{23} = H'_x \; F^{14} = -E'_x$

$\{$

(64) $\{ F^{31} = H'_y \; F^{24} = -E'_y$

$\{$

$\{ F^{12} = H'_z \; F^{34} = -E'_z$

which quantities become equal to H_x ... E_x in the case of the special relativity theory, and besides

$J^1 = i_x \ldots J^4 = \varrho$

we get instead of (63)

{ rot H' - $\partial E'/\partial t = i$

(63a) {

{ div E' = ϱ

The equations (60), (62) and (63) give thus a generalisation of Maxwell's field-equations in vacuum, which remains true in our chosen system of co-ordinates.

The energy-components of the electro-magnetic field.

Let us form the inner-product

(65) $K_\sigma = F_{\sigma\mu} J^\mu$.

According to (61) its components can be written down in the three-dimensional notation.

{ $K_1 = \varrho E_x + [i, H]_x$

(65a) { ———

{ $K_4 =$ — (i, E).

K_σ is a covariant four-vector whose components are equal to the negative impulse and energy which are transferred to the electro-magnetic field per unit of time, and per unit of volume, by the electrical masses. If the electrical masses be free, that is, under the influence of the electro-magnetic field only, then the covariant four-vector K_σ will vanish.

In order to get the energy components T_σ^ν of the electro-magnetic field, we require only to give to the equation $K_\sigma = 0$, the form of the equation (57).

From (63) and (65) we get first,

$K_\sigma = F_{\sigma\mu}\, \partial F_{\mu\nu}/\partial x_\nu$

$= \partial/\partial x_\nu\, (F_{\sigma\mu} F^{\mu\nu}) - F^{\mu\nu}\, \partial F_{\sigma\mu}/\partial x_\nu$.

On account of (60) the second member on the right-hand side admits of the transformation—

$F^{\mu\nu}\, \partial F_{\sigma\mu}/\partial x_\nu = -\frac{1}{2} F^{\mu\nu}\, \partial F_{\mu\nu}/\partial x_\sigma$

$= -\frac{1}{2} g^{\mu\alpha} g^{\nu\beta} F_{\alpha\beta}\, \partial F_{\mu\nu}/\partial x_\sigma$.

Owing to symmetry, this expression can also be written in the form

$$= -1/4\ [g^{\mu\alpha} g^{\nu\beta} F_{\alpha\beta}\ \partial F_{\mu\nu}/\partial x_\sigma$$

$$+ g^{\mu\alpha} g^{\nu\beta}\ \partial F_{\alpha\beta}/\partial x_\sigma\ F_{\mu\nu}],$$

which can also be put in the form

$$- 1/4\ \partial/\partial x_\sigma\ (g^{\mu\alpha} g^{\nu\beta} F_{\alpha\beta} F_{\mu\nu})$$

$$+ 1/4\ F_{\alpha\beta} F_{\mu\nu}\ \partial/\partial x_\sigma\ (g^{\mu\alpha} g^{\nu\beta}).$$

The first of these terms can be written shortly as

$$- 1/4\ \partial/\partial x_\sigma\ (F^{\mu\nu} F_{\mu\nu}),$$

and the second after differentiation can be transformed in the form

$$- ½\ F^{\mu\tau} F_{\mu\nu}\ g^{\nu\varrho}\ \partial g_{\sigma\tau}/\partial x_\sigma.$$

If we take all the three terms together, we get the relation

$$(66)\ K_\sigma = \partial \tau_\sigma{}^\nu/\partial x_\nu - ½\ g^{\tau\mu}\ \partial g_{\mu\nu}/\partial x_\sigma\ \tau_\tau{}^\nu$$

where

$$(66a)\ \tau_\sigma{}^\nu = -F_{\sigma\alpha} F^{\nu\alpha} + 1/4\ \delta_\sigma{}^\nu F_{\alpha\beta} F^{\alpha\beta}.$$

On account of (30) the equation (66) becomes equivalent to (57) and (57a) when K_σ vanishes. Thus $\tau_\sigma{}^\nu$'s are the energy-components of the electro-magnetic field. With the help of (61) and (64) we can easily show that the energy-components of the electro-magnetic field, in the case of the special relativity theory, give rise to the well-known Maxwell-Poynting expressions.

We have now deduced the most general laws which the gravitation-field and matter satisfy when we use a co-ordinate system for which $\sqrt{(-g)} = 1$. Thereby we achieve an important simplification in all our formulas and calculations, without renouncing the conditions of general covariance, as we have obtained the equations through a specialisation of the co-ordinate system from the general covariant-equations. Still the question is not without formal interest, whether, when the energy-components of the gravitation-field and matter is defined in a generalised manner without any specialisation of co-ordinates, the laws of conservation have the form of the equation (56), and the field-equations of gravitation hold in the form

(52) or (52a); such that on the left-hand side, we have a divergence in the usual sense, and on the right-hand side, the sum of the energy-components of matter and gravitation. I have found out that this is indeed the case. But I am of opinion that the communication of my rather comprehensive work on this subject will not pay, for nothing essentially new comes out of it.

E. §21. Newton's theory as a first approximation.

We have already mentioned several times that the special relativity theory is to be looked upon as a special case of the general, in which $g_{\mu\nu}$'s have constant values (4). This signifies, according to what has been said before, a total neglect of the influence of gravitation. We get one important approximation if we consider the case when $g_{\mu\nu}$'s differ from (4) only by small magnitudes (compared to 1) where we can neglect small quantities of the second and higher orders (first aspect of the approximation.)

Further it should be assumed that within the space-time region considered, $g_{\mu\nu}$'s at infinite distances (using the word infinite in a spatial sense) can, by a suitable choice of co-ordinates, tend to the limiting values (4); *i.e.*, we consider only those gravitational fields which can be regarded as produced by masses distributed over finite regions.

We can assume that this approximation should lead to Newton's theory. For it however, it is necessary to treat the fundamental equations from another point of view. Let us consider the motion of a particle according to the equation (46). In the case of the special relativity theory, the components

$$dx_1/ds,\ dx_2/ds,\ dx_3/ds,$$

can take any values. This signifies that any velocity

$$v = \sqrt{((dx_1/dx_4)^2 + (dx_2/dx_4)^2 + (dx_3/dx_4)^2)}$$

can appear which is less than the velocity of light in vacuum ($v < 1$). If we finally limit ourselves to the consideration of the case when v is small compared to the velocity of light, it signifies that the components

$$dx_1/ds,\ dx_2/ds,\ dx_3/ds,$$

can be treated as small quantities, whereas dx_4/ds is equal to 1, up to the second-order magnitudes (the second point of view for approximation).

Now we see that, according to the first view of approximation, the magnitudes $\gamma_{\mu\nu}^{\tau}$'s are all small quantities of at least the first order. A glance at (46) will also show, that in this equation according to the second view of approximation, we are only to take into account those terms for which $\mu = \nu = 4$.

By limiting ourselves only to terms of the lowest order we get instead of (46), first, the equations:—

$d^2x_\tau/dt^2 = \Gamma_{44}^\tau$, where $ds = dx_4 = dt$,

or by limiting ourselves only to those terms which according to the first stand-point are approximations of the first order,

It must be admitted, that this introduction of the energy-tensor of matter cannot be justified by means of the Relativity-Postulate alone; for we have in the foregoing analysis deduced it from the condition that the energy of the gravitation-field should exert gravitating action in the same way as every other kind of energy. The strongest ground for the choice of the above equation however lies in this, that they lead, as their consequences, to equations expressing the conservation of the components of total energy (the impulses and the energy) which exactly correspond to the equations (49) and (49a). This shall be shown afterwards.

§17. The laws of conservation in the general case.

The equations (52) can be easily so transformed that the second member on the right-hand side vanishes. We reduce (52) with reference to the indices μ and σ and subtract the equation so obtained after multiplication with $\frac{1}{2}\,\delta_\mu^\sigma$ from (52).

We obtain,

(52a) $\partial/\partial x_\alpha(g^{\sigma\beta}\,\Gamma_{\mu\beta}^\alpha - \frac{1}{2}\,\delta_\mu^\sigma\,g^{\lambda\beta}\,\Gamma_{\lambda\beta}^\alpha)$

$= -\varkappa(t_\mu^\sigma + T_\mu^\sigma)$

we operate on it by $\partial/\partial x_\sigma$. Now,

$\partial^2/\partial x_\alpha \partial x_\sigma\,(g^{\sigma\beta}\Gamma_{\mu\beta}^\alpha)$

$= -\frac{1}{2}\,\partial^2/\partial x_\alpha \partial x_\sigma\,[g^{\sigma\beta}\,g^{\alpha\lambda}(\partial g_{\mu\lambda}/\partial x_\beta$

$+ \partial g_{\beta\lambda}/\partial x_\mu - \partial g_{\mu\beta}/\partial x_\lambda)]$.

The first and the third member of the round bracket lead to expressions which cancel one another, as can be easily seen by interchanging the summation-indices α, and σ, on the one hand, and β and λ, on the other.

The second term can be transformed according to (31). So that we get,

(54) $\partial^2/\partial x_\alpha \partial x_\sigma\,(g^{\sigma\beta}\gamma_{\mu\beta}^\alpha)$

$= \frac{1}{2}\,\partial^3 g^{\alpha\beta}/\partial x_\sigma \partial x_\beta \partial x_\mu$

The second member of the expression on the left-hand side of (52a) leads first to

$- \frac{1}{2} \partial^2/\partial x_\alpha \partial x_\mu (g^{\lambda\beta}\Gamma_{\lambda\beta}{}^\alpha)$ or

to $1/4\ \partial^2/\partial x_\alpha \partial x_\mu [g^{\lambda\beta} g^{\alpha\delta}(\partial g_{\delta\lambda}/\partial x_\beta$
$+ \partial g_{\delta\beta}/\partial x_\lambda - \partial g_{\lambda\beta}/\partial x_\delta)]$.

The expression arising out of the last member within the round bracket vanishes according to (29) on account of the choice of axes. The two others can be taken together and give us on account of (31), the expression

$-\frac{1}{2} \partial^3 g^{\alpha\beta}/\partial x_\alpha \partial x_\beta \partial x_\mu$

So that remembering (54) we have

(55) $\partial^2/\partial x_\alpha \partial x_\sigma (g^{\sigma\beta}\Gamma_{\mu\beta}{}^\alpha$
$- \frac{1}{2} \delta_\mu{}^\sigma g^{\lambda\beta} \Gamma_{\lambda\beta}{}^\alpha) = 0.$

identically.

From (55) and (52a) it follows that

(56) $\partial/\partial x_\sigma (t_\mu{}^\sigma + T_\mu{}^\sigma) = 0$

From the field equations of gravitation, it also follows that the conservation-laws of impulse and energy are satisfied. We see it most simply following the same reasoning which lead to equations (49a); only instead of the energy-components of the gravitational-field, we are to introduce the total energy-components of matter and gravitational field.

§18. The Impulse-energy law for matter as a consequence of the field-equations.

If we multiply (53) with $\partial g^{\mu\nu}/\partial x_\sigma$, we get in a way similar to §15, remembering that

$g_{\mu\nu} \partial g^{\mu\nu}/\partial x_\sigma$ vanishes,

the equations $\partial t_\sigma{}^\alpha/\partial x_\alpha - \frac{1}{2} \partial g^{\mu\nu}/\partial x_\sigma T_{\mu\nu} = 0$

or remembering (56)

(57) $\partial T_\sigma{}^\alpha/\partial x_\alpha + \frac{1}{2} \partial g^{\mu\nu}/\partial x_\sigma T_{\mu\nu} = 0$

A comparison with (41b) shows that these equations for the above choice of co-ordinates ($\sqrt{(-g)} = 1$) asserts nothing but the vanishing of the divergence of the tensor of the energy-components of matter.

Physically the appearance of the second term on the left-hand side shows that for matter alone the law of conservation of impulse and energy cannot hold; or can only hold when $g^{\mu\nu}$'s are constants; *i.e.*, when the field of gravitation vanishes. The second member is an expression for impulse and energy which the gravitation-field exerts per time and per volume upon matter. This comes out clearer when instead of (57) we write it in the form of (47).

(57a) $\partial T_\sigma^\alpha / \partial x_\alpha = -\Gamma_{\sigma\beta}^\alpha T_\alpha^\beta$.

The right-hand side expresses the interaction of the energy of the gravitational-field on matter. The field-equations of gravitation contain thus at the same time 4 conditions which are to be satisfied by all material phenomena. We get the equations of the material phenomena completely when the latter is characterised by four other differential equations independent of one another.

D. THE "MATERIAL" PHENOMENA.

The Mathematical auxiliaries developed under 'B' at once enables us to generalise, according to the generalised theory of relativity, the physical laws of matter (Hydrodynamics, Maxwell's Electro-dynamics) as they lie already formulated according to the special-relativity-theory. The generalised Relativity Principle leads us to no further limitation of possibilities; but it enables us to know exactly the influence of gravitation on all processes without the introduction of any new hypothesis.

It is owing to this, that as regards the physical nature of matter (in a narrow sense) no definite necessary assumptions are to be introduced. The question may lie open whether the theories of the electro-magnetic field and the gravitational-field together, will form a sufficient basis for the theory of matter. The general relativity postulate can teach us no new principle. But by building up the theory it must be shown whether electro-magnetism and gravitation together can achieve what the former alone did not succeed in doing.

§19. Euler's equations for frictionless adiabatic liquid.

Let p and ϱ, be two scalars, of which the first denotes the pressure and the last the density of the fluid; between them there is a relation. Let the contravariant symmetrical tensor

$$T^{\alpha\beta} = -g^{\alpha\beta} p + \varrho \, dx_\alpha/ds \, dx_\beta/ds \quad (58)$$

be the contra-variant energy-tensor of the liquid. To it also belongs the covariant tensor

$$(58a) \quad T_{\mu\nu} = -g_{\mu\nu} p + g_{\mu\alpha} \, dx_\alpha/ds \, g_{\mu\beta} \, dx_\beta/ds \, \varrho$$

as well as the mixed tensor

$$(58b) \quad T^\alpha_\sigma = -\delta^\alpha_\sigma p + g_{\sigma\beta} \, dx_\beta/ds \, dx_\alpha/ds \, \varrho.$$

If we put the right-hand side of (58b) in (57a) we get the general hydrodynamical equations of Euler according to the generalised relativity theory. This in principle completely solves the problem of motion; for the four equations (57a) together with the given equation between p and ϱ, and the equation

$$g_{\alpha\beta} \, dx\_\alpha/ds \, dx_\beta/ds = 1,$$

are sufficient, with the given values of $g_{\alpha\beta}$, for finding out the six unknowns

p, ϱ, dx_1/ds, dx_2/ds, dx_3/ds dx_4/ds.

If $g_{\mu\nu}$'s are unknown we have also to take the equations (53). There are now 11 equations for finding out 10 functions g, so that the number is more than sufficient. Now it is be noticed that the equation (57a) is already contained in (53), so that the latter only represents (7) independent equations. This indefiniteness is due to the wide freedom in the choice of co-ordinates, so that mathematically the problem is indefinite in the sense that three of the space-functions can be arbitrarily chosen.

§20. Maxwell's Electro-Magnetic field-equations.

Let φ_ν be the components of a covariant four-vector, the electro-magnetic potential; from it let us form according to (36) the components $F_{\varrho\sigma}$ of the covariant six-vector of the electro-magnetic field according to the system of equations

(59) $F_{\varrho\sigma} = \partial\varphi_\varrho/\partial x_\sigma - \partial\varphi_\sigma/\partial x_\varrho$.

From (59), it follows that the system of equations

(60) $\partial F_{\varrho\sigma}/\partial x_\tau + \partial F_{\sigma\tau}/\partial x_\varrho + \partial F_{\tau\varrho}/\partial x_\sigma = 0$

is satisfied of which the left-hand side, according to (37), is an anti-symmetrical tensor of the third kind. This system (60) contains essentially four equations, which can be thus written:—

{ $\partial F_{23}/\partial x_4 + \partial F_{34}/\partial x_2\ \partial F_{42}/\partial x_3 = 0$

{

{ $\partial F_{34}/\partial x_1 + \partial F_{41}/\partial x_3\ \partial F_{13}/\partial x_4 = 0$

(60a) {

{ $\partial F_{41}/\partial x_2 + \partial F_{12}/\partial x_4\ \partial F_{24}/\partial x_1 = 0$

{

{ $\partial F_{12}/\partial x_3 + \partial F_{23}/\partial x_1\ \partial F_{31}/\partial x_2 = 0$.

This system of equations corresponds to the second system of equations of Maxwell. We see it at once if we put

{ $F_{23} = H_x$ $F_{14} = E_x$

{

$$(61) \begin{cases} F_{31} = H_y & F_{24} = E_y \\ F_{12} = H_z & F_{34} = E_z \end{cases}$$

Instead of (60a) we can therefore write according to the usual notation of three-dimensional vector-analysis:—

$$(60b) \begin{cases} \partial H/\partial t + \text{rot } E = 0 \\ \text{div } H = 0. \end{cases}$$

The first Maxwellian system is obtained by a generalisation of the form given by Minkowski.

We introduce the contra-variant six-vector $F_{\alpha\beta}$ by the equation

$$(62) \quad F^{\mu\nu} = g^{\mu\alpha} g^{\nu\beta} F_{\alpha\beta},$$

and also a contra-variant four-vector J^{μ}, which is the electrical current-density in vacuum. Then remembering (40) we can establish the system of equations, which remains invariant for any substitution with determinant 1 (according to our choice of co-ordinates).

$$(63) \quad \partial F^{\mu\nu}/\partial x_{\nu} = J^{\mu}$$

If we put

$$(64) \begin{cases} F^{23} = H'_x & F^{14} = -E'_x \\ F^{31} = H'_y & F^{24} = -E'_y \\ F^{12} = H'_z & F^{34} = -E'_z \end{cases}$$

which quantities become equal to H_x ... E_x in the case of the special relativity theory, and besides

$$J^1 = i_x \ldots J^4 = \varrho$$

we get instead of (63)

$\{$ rot H' - $\partial E'/\partial t = i$

(63a) $\{$

$\{$ div E' = ϱ

The equations (60), (62) and (63) give thus a generalisation of Maxwell's field-equations in vacuum, which remains true in our chosen system of co-ordinates.

The energy-components of the electro-magnetic field.

Let us form the inner-product

(65) $K_\sigma = F_{\sigma\mu} J^\mu$.

According to (61) its components can be written down in the three-dimensional notation.

$\{ K_1 = \varrho E_x + [i, H]_x$

(65a) { — — —

$\{ K_4 = — (i, E)$.

K_σ is a covariant four-vector whose components are equal to the negative impulse and energy which are transferred to the electro-magnetic field per unit of time, and per unit of volume, by the electrical masses. If the electrical masses be free, that is, under the influence of the electro-magnetic field only, then the covariant four-vector K_σ will vanish.

In order to get the energy components T_σ^ν of the electro-magnetic field, we require only to give to the equation $K_\sigma = 0$, the form of the equation (57).

From (63) and (65) we get first,

$K_\sigma = F_{\sigma\mu} \partial F_{\mu\nu}/\partial x_\nu$

$= \partial/\partial x_\nu (F_{\sigma\mu} F^{\mu\nu}) - F^{\mu\nu} \partial F_{\sigma\mu}/\partial x_\nu$.

On account of (60) the second member on the right-hand side admits of the transformation—

$F^{\mu\nu} \partial F_{\sigma\mu}/\partial x_\nu = -\frac{1}{2} F^{\mu\nu} \partial F_{\mu\nu}/\partial x_\sigma$

$= -\frac{1}{2} g^{\mu\alpha} g^{\nu\beta} F_{\alpha\beta} \partial F_{\mu\nu}/\partial x_\sigma$.

Owing to symmetry, this expression can also be written in the form

$$= -1/4 \, [g^{\mu\alpha} g^{\nu\beta} F_{\alpha\beta} \, \partial F_{\mu\nu}/\partial x_\sigma$$

$$+ g^{\mu\alpha} g^{\nu\beta} \, \partial F_{\alpha\beta}/\partial x_\sigma \, F_{\mu\nu}],$$

which can also be put in the form

$$- 1/4 \, \partial/\partial x_\sigma \, (g^{\mu\alpha} g^{\nu\beta} F_{\alpha\beta} F_{\mu\nu})$$

$$+ 1/4 \, F_{\alpha\beta} F_{\mu\nu} \, \partial/\partial x_\sigma \, (g^{\mu\alpha} g^{\nu\beta}).$$

The first of these terms can be written shortly as

$$- 1/4 \, \partial/\partial x_\sigma \, (F^{\mu\nu} F_{\mu\nu}),$$

and the second after differentiation can be transformed in the form

$$- ½ \, F^{\mu\tau} F_{\mu\nu} \, g^{\nu\varrho} \, \partial g_{\sigma\tau}/\partial x_\sigma.$$

If we take all the three terms together, we get the relation

$$(66) \; K_\sigma = \partial \tau_\sigma^\nu/\partial x_\nu - ½ \, g^{\tau\mu} \, \partial g_{\mu\nu}/\partial x_\sigma \, \tau_\tau^\nu$$

where

$$(66a) \; \tau_\sigma^\nu = -F_{\sigma\alpha} F^{\nu\alpha} + 1/4 \, \delta_\sigma^\nu F_{\alpha\beta} F^{\alpha\beta}.$$

On account of (30) the equation (66) becomes equivalent to (57) and (57a) when K_σ vanishes. Thus τ_σ^ν's are the energy-components of the electro-magnetic field. With the help of (61) and (64) we can easily show that the energy-components of the electro-magnetic field, in the case of the special relativity theory, give rise to the well-known Maxwell-Poynting expressions.

We have now deduced the most general laws which the gravitation-field and matter satisfy when we use a co-ordinate system for which $\sqrt{(-g)} = 1$. Thereby we achieve an important simplification in all our formulas and calculations, without renouncing the conditions of general covariance, as we have obtained the equations through a specialisation of the co-ordinate system from the general covariant-equations. Still the question is not without formal interest, whether, when the energy-components of the gravitation-field and matter is defined in a generalised manner without any specialisation of co-ordinates, the laws of conservation have the form of the equation (56), and the field-equations of gravitation hold in the form

(52) or (52a); such that on the left-hand side, we have a divergence in the usual sense, and on the right-hand side, the sum of the energy-components of matter and gravitation. I have found out that this is indeed the case. But I am of opinion that the communication of my rather comprehensive work on this subject will not pay, for nothing essentially new comes out of it.

E. §21. Newton's theory as a first approximation.

We have already mentioned several times that the special relativity theory is to be looked upon as a special case of the general, in which $g_{\mu\nu}$'s have constant values (4). This signifies, according to what has been said before, a total neglect of the influence of gravitation. We get one important approximation if we consider the case when $g_{\mu\nu}$'s differ from (4) only by small magnitudes (compared to 1) where we can neglect small quantities of the second and higher orders (first aspect of the approximation.)

Further it should be assumed that within the space-time region considered, $g_{\mu\nu}$'s at infinite distances (using the word infinite in a spatial sense) can, by a suitable choice of co-ordinates, tend to the limiting values (4); *i.e.*, we consider only those gravitational fields which can be regarded as produced by masses distributed over finite regions.

We can assume that this approximation should lead to Newton's theory. For it however, it is necessary to treat the fundamental equations from another point of view. Let us consider the motion of a particle according to the equation (46). In the case of the special relativity theory, the components

$$dx_1/ds, dx_2/ds, dx_3/ds,$$

can take any values. This signifies that any velocity

$$v = \sqrt{((dx_1/dx_4)^2 + (dx_2/dx_4)^2 + (dx_3/dx_4)^2)}$$

can appear which is less than the velocity of light in vacuum ($v < 1$). If we finally limit ourselves to the consideration of the case when v is small compared to the velocity of light, it signifies that the components

$$dx_1/ds, dx_2/ds, dx_3/ds,$$

can be treated as small quantities, whereas dx_4/ds is equal to 1, up to the second-order magnitudes (the second point of view for approximation).

Now we see that, according to the first view of approximation, the magnitudes $\gamma_{\mu\nu}^{\tau}$'s are all small quantities of at least the first order. A glance at (46) will also show, that in this equation according to the second view of approximation, we are only to take into account those terms for which $\mu = \nu = 4$.

By limiting ourselves only to terms of the lowest order we get instead of (46), first, the equations:—

$d^2x_\tau/dt^2 = \Gamma_{44}{}^\tau$, where $ds = dx_4 = dt$,

or by limiting ourselves only to those terms which according to the first stand-point are approximations of the first order,

$$\frac{d^2 x_\tau}{dt^2} = \begin{bmatrix} 44 \\ \tau \end{bmatrix} \quad (\tau = 1, 2, 3)$$

$$\frac{d^2 x_4}{dt^2} = -\begin{bmatrix} 44 \\ 4 \end{bmatrix}.$$

If we further assume that the gravitation-field is quasi-static, *i.e.*, it is limited only to the case when the matter producing the gravitation-field is moving slowly (relative to the velocity of light) we can neglect the differentiations of the positional co-ordinates on the right-hand side with respect to time, so that we get

(67) $d^2x_\tau/dt^2 = -\frac{1}{2}\, \partial g_{44}/\partial x_\tau$ (τ, = 1, 2, 3)

This is the equation of motion of a material point according to Newton's theory, where $g_{44}/2$ plays the part of gravitational potential. The remarkable thing in the result is that in the first-approximation of motion of the material point, only the component g_{44} of the fundamental tensor appears.

Let us now turn to the field-equation (53). In this case, we have to remember that the energy-tensor of matter is exclusively defined in a narrow sense by the density ϱ of matter, *i.e.*, by the second member on the right-hand side of 58 [(58a, or 58b)]. If we make the necessary approximations, then all component vanish except

$\tau_{44} = \varrho = \tau$.

On the left-hand side of (53) the second term is an infinitesimal of the second order, so that the first leads to the following terms in the approximation, which are rather interesting for us:

$$\frac{\partial}{\partial x_1}\begin{bmatrix}\mu\nu\\1\end{bmatrix}+\frac{\partial}{\partial x_2}\begin{bmatrix}\mu\nu\\2\end{bmatrix}+\frac{\partial}{\partial x_3}\begin{bmatrix}\mu\nu\\3\end{bmatrix}-\frac{\partial}{\partial x_4}\begin{bmatrix}\mu\nu\\4\end{bmatrix}.$$

By neglecting all differentiations with regard to time, this leads, when $\mu = \nu = 4$, to the expression

$$-\frac{1}{2}\left(\frac{\partial^2 g_{44}}{\partial x_1^2}+\frac{\partial^2 g_{44}}{\partial x_2^2}+\frac{\partial^2 g_{44}}{\partial x_3^2}\right)=-\frac{1}{2}\nabla^2 g_{44}.$$

The last of the equations (53) thus leads to

(68) $\nabla^2 g_{44} = \varkappa\varrho$.

The equations (67) and (68) together, are equivalent to Newton's law of gravitation.

For the gravitation-potential we get from (67) and (68) the exp.

(68a.) $-\varkappa/(8\pi)\int \varrho d\tau/r$

whereas the Newtonian theory for the chosen unit of time gives

$-K/c^2 \int \varrho d\tau/r$,

where K denotes usually the gravitation-constant. 6.7×10^{-8}; equating them we get

(69) $\varkappa = 8\pi K/c^2 = 1.87 \times 10^{-27}$.

§22. Behaviour of measuring rods and clocks in a statical gravitation-field. Curvature of light-rays. Perihelion-motion of the paths of the Planets.

In order to obtain Newton's theory as a first approximation we had to calculate only g_{44}, out of the 10 components $g_{\mu\nu}$ of the gravitation-potential, for that is the only component which comes in the first approximate equations of motion of a material point in a gravitational field.

We see however, that the other components of $g_{\mu\nu}$ should also differ from the values given in (4) as required by the condition $g = -1$.

For a heavy particle at the origin of co-ordinates and generating the gravitational field, we get as a first approximation the symmetrical solution of the equation:—

$\{ g_{\varrho\sigma} = -\delta_{\varrho\sigma} - \alpha(x_\varrho x_\sigma)/r^3$ (ϱ and σ 1, 2, 3)

{

(70) $\{ g_{\varrho 4} = g_{4\varrho} = 0$ (ϱ 1, 2, 3)

{

$\{ g_{44} = 1 - \alpha/r.$

$\delta_{\varrho\sigma}$ is 1 or 0, according as $\varrho = \sigma$ or not and r is the quantity

$+\sqrt{(x_1^2 + x_2^2 + x_3^2)}$.

On account of (68a) we have

(70a) $\alpha = \varkappa M/4\pi$

where M denotes the mass generating the field. It is easy to verify that this solution satisfies approximately the field-equation outside the mass M.

Let us now investigate the influences which the field of mass M will have upon the metrical properties of the field. Between the lengths and times measured locally on the one hand, and the differences in co-ordinates dx_ν on the other, we have the relation

$ds^2 = g_{\mu\nu}\, dx_\mu\, dx_\nu$.

For a unit measuring rod, for example, placed parallel to the x axis, we have to put

$ds^2 = -1,\ dx_2 = dx_3 = dx_4 = 0$

then $-1 = g_{11} dx_1^2$.

If the unit measuring rod lies on the x axis, the first of the equations (70) gives

$g_{11} = -(1 + \alpha/r)$.

From both these relations it follows as a first approximation that

(71) $dx = 1 - \alpha/2r$.

The unit measuring rod appears, when referred to the co-ordinate-system, shortened by the calculated magnitude through the presence of the gravitational field, when we place it radially in the field.

Similarly we can get its co-ordinate-length in a tangential position, if we put for example

$ds^2 = -1,\ dx_1 = dx_3 = dx_4 = 0,\ x_1 = r,\ x_2 = x_3 = 0$

we then get

(71a) $-1 = g_{22}\, dx_2^2 = -dx_2^2$.

The gravitational field has no influence upon the length of the rod, when we put it tangentially in the field.

Thus Euclidean geometry does not hold in the gravitational field even in the first approximation, if we conceive that one and the same rod independent of its position and its orientation can serve as the measure of the same extension. But a glance at (70a) and (69) shows that the expected difference is much too small to be noticeable in the measurement of earth's surface.

We would further investigate the rate of going of a unit-clock which is placed in a statical gravitational field. Here we have for a period of the clock

$ds = 1,\ dx_1 = dx_2\ dx_3 = 0;$

then we have

$1 = g_{44} dx_4^2$

$dx_4 = 1/\sqrt{(g_{44})} = 1/\sqrt{(1 + (g_{44} - 1))} = 1 - (g_{44} - 1)/2$

or $dx_4 = 1 + k/8\pi \int \varrho d\tau/r$.

Therefore the clock goes slowly what it is placed in the neighbourhood of ponderable masses. It follows from this that the spectral lines in the light coming to us from the surfaces of big stars should appear shifted towards the red end of the spectrum.

Let us further investigate the path of light-rays in a statical gravitational field. According to the special relativity theory, the velocity of light is given by the equation

$$-dx_1^2 - dx_2^2 - dx_3^2 + dx_4^2 = 0;$$

thus also according to the generalised relativity theory it is given by the equation

$$(73)\ ds^2 = g_{\mu\nu}\, dx_\mu\, dx_\nu = 0.$$

If the direction, *i.e.*, the ratio $dx_1 : dx_2 : dx_3$ is given, the equation (73) gives the magnitudes

$$dx_1/dx_4,\ dx_2/dx_4,\ dx_3/dx_4,$$

and with it the velocity,

$$\sqrt{((dx_1/dx_4)^2 + (dx_2/dx_4)^2 + (dx_3/dx_4)^2)} = \gamma,$$

in the sense of the Euclidean Geometry. We can easily see that, with reference to the co-ordinate system, the rays of light must appear curved in case $g_{\mu\nu}$'s are not constants. If n be the direction perpendicular to the direction of propagation, we have, from Huygen's principle, that light-rays (taken in the plane (γ, n)] must suffer a curvature $\partial\lambda/\partial n$.

Let us find out the curvature which a light-ray suffers when it goes by a mass M at a distance Δ from it. If we use the co-ordinate system according to the above scheme, then the total bending B of light-rays (reckoned positive when it is concave to the origin) is given as a sufficient approximation by

$$B = \int_{-\infty}^{\infty} \partial\gamma/\partial[x]_1\, dx_2$$

where (73) and (70) gives

$$\gamma = \sqrt{(-g_{44}/g_{22})} = 1 - \alpha/2r\,(1 + x_2^2/r^2).$$

The calculation gives

$$B = 2\alpha/\Delta = KM/2\pi\Delta.$$

A ray of light just grazing the sun would suffer a bending of 1·7", whereas one coming by Jupiter would have a deviation of about ·02".

If we calculate the gravitation-field to a greater order of approximation and with it the corresponding path of a material particle of a relatively small (infinitesimal) mass we get a deviation of the following kind from the Kepler-Newtonian Laws of Planetary motion. The Ellipse of Planetary motion suffers a slow rotation in the direction of motion, of amount

(75) $s = 24\pi^3a^2/\tau^2c^2(1 - e^2)$ per revolution.

In this Formula 'a' signifies the semi-major axis, c, the velocity of light, measured in the usual way, e, the eccentricity, τ, the time of revolution in seconds.

The calculation gives for the planet Mercury, a rotation of path of amount 43" per century, corresponding sufficiently to what has been found by astronomers (Leverrier). They found a residual perihelion motion of this planet of the given magnitude which can not be explained by the perturbation of the other planets.

NOTES

Note 1.

The fundamental electro-magnetic equations of Maxwell for stationary media are:—

$\text{curl } H = 1/c\,(\partial D/\partial t + \varrho v)$ (1)

$\text{curl } E = -1/c\,\partial B/\partial t$ (2)

$\text{div } D = \varrho$

$B = \mu H$

$\text{div } B = 0$

$D = kE$

According to Hertz and Heaviside, these require modification in the case of moving bodies.

Now it is known that due to motion alone there is a change in a vector *R* given by

$(\partial R/\partial t)$ due to motion $= u.\ \text{div } R + \text{curl } [Ru]$

where u is the vector velocity of the moving body and $[Ru]$ the vector product of R and u.

Hence equations (1) and (2) become

$c\ \text{curl } H = \partial D/\partial t + u\ \text{div } D + \text{curl Vect. } [Du] + \varrho v$ (1·1)

and

$-c\ \text{curl } E = \partial B/\partial t + u\ \text{div } B + \text{curl Vect. } [Bu]$ (2·1)

which gives finally, for $\varrho = 0$ and $\text{div } B = 0$,

$\partial D/\partial t + u\ \text{div } D = c\ \text{curl } (H - 1/c\ \text{Vect. } [Du])$ (1·2)

$\partial B/\partial t = -c\ \text{curl } (E - 1/c\ \text{Vect. } [uB])$ (2·2)

Let us consider a beam travelling along the x-axis, with apparent velocity v (*i.e.*, velocity with respect to the fixed ether) in medium moving with velocity $u_x = u$ in the same direction.

Then if the electric and magnetic vectors are proportional to $e^{iA(x - vt)}$, we have

$\partial/\partial x = iA$, $\partial/\partial t = -iAv$, $\partial/\partial y = \partial/\partial z = 0$, $u_y = u_z = 0$

Then $\partial D\_y/\partial t = -c\partial H_z/\partial x - u\partial D_y/\partial z$ ... (1·21)

and $\partial B_z/\partial t = -c\partial E_y/\partial x - u\partial B_z/\partial x$ (2·21)

Since D = KE and B = μH, we have

$iAv(\kappa Ey) = -ciA(H_z + uKE_y)$ (1·22)

$iAv(\mu H_z) = -ciA(E_y + u\mu H_z)$ (2·22)

or $v(K - u)E_y = cH_z$ (1·23)

$\mu(v - u)H_z = cE_y$ (2·23)

Multiplying (1·23) by (2·23)

$\mu K(v - u)^2 = c^2$

Hence $(v - u)^2 = c^2/\mu k = v_0^2$

$\therefore v = v_0 + u,$

making Fresnelian convection co-efficient simply unity.

Equations (1·21) and (2·21) may be obtained more simply from physical considerations.

According to Heaviside and Hertz, the real seat of both electric and magnetic polarisation is the moving medium itself. Now at a point which is fixed with respect to the ether, the rate of change of electric polarisation is $\delta D/\delta t$.

Consider a slab of matter moving with velocity u_x along the x-axis, then even in a stationary field of electrostatic polarisation, that is, for a field in which $\delta D/\delta t = 0$, there will be some change in the polarisation of the body due to its motion, given by $u_x(\delta D/\delta x)$. Hence we must add this term to a purely temporal rate of change $\delta D/\delta t$. Doing this we immediately arrive at equations (1·21) and (2·21) for the special case considered there.

Thus the Hertz-Heaviside form of field equations gives *unity* as the value for the Fresnelian convection co-efficient. It has been shown in the historical introduction how this is entirely at variance with the observed optical facts. As a matter of fact, Larmor has shown (Aether and Matter) that $1 - 1/\mu^2$ is not only sufficient but is also necessary, in order to explain experiments of the Arago prism type.

A short summary of the electromagnetic experiments bearing on this question, has already been given in the introduction.

According to Hertz and Heaviside the total polarisation is situated in the medium itself and is completely carried away by it. Thus the electromagnetic effect outside a moving medium should be proportional to K, the specific inductive capacity.

Rowland showed in 1876 that when a charged condenser is rapidly rotated (the dielectric remaining stationary), the magnetic effect outside is proportional to K, the Sp. Ind. Cap.

Röntgen (Annalen der Physik 1888, 1890) found that if the dielectric is rotated while the condenser remains stationary, the effect is proportional to $K - 1$.

Eichenwald (Annalen der Physik 1903, 1904) rotated together both condenser and dielectric and found that the magnetic effect was proportional to the potential difference and to the angular velocity, but was completely independent of K. This is of course quite consistent with Rowland and Röntgen.

Blondlot (Comptes Rendus, 1901) passed a current of air in a steady magnetic field H_y, ($H = H_z = 0$). If this current of air moves with velocity u_x along the x-axis, an electromotive force would be set up along the z-axis, due to the relative motion of matter and magnetic tubes of induction. A pair of plates at $z = \pm a$, will be charged up with density $\varrho = D_z = KE = K. u_s H_y/c$. But Blondlot failed to detect any such effect.

H. A. Wilson (Phil. Trans. Royal Soc. 1904) repeated the experiment with a cylindrical condenser made of ebony, rotating in a magnetic field parallel to its own axis. He observed a change proportional to $K — 1$ and not to K.

Thus the above set of electro-magnetic experiments contradict the Hertz-Heaviside equations, and these must be abandoned.

[P. C. M.]

Note 2.
Lorentz Transformation.

Lorentz. Versuch einer theorie der elektrischen und optischen Erscheinungen im bewegten Körpern.

(Leiden—1895).

Lorentz. Theory of Electrons (English edition), pages 197-200, 230, also notes 73, 86, pages 318, 328.

Lorentz wanted to explain the Michelson-Morley null-effect. In order to do so, it was obviously necessary to explain the Fitzgerald contraction. Lorentz worked on the hypothesis that an electron itself undergoes contraction when moving. He introduced new variables for the moving system defined by the following set of equations.

$$x^1 = \beta(x - ut), y^1 = y, z^1 = z, t^1 = \beta(t - (u/c^2)\cdot x)$$

and for velocities, used

$$v_x^1 = \beta^2 v_x + u,\ v_y^1 = \beta v_y,\ v_z^1 = \beta v_z \text{ and } \varrho^1 = \varrho/\beta.$$

With the help of the above set of equations, which is known as the Lorentz transformation, he succeeded in showing how the Fitzgerald contraction results as a consequence of "fortuitous compensation of opposing effects."

It should be observed that the Lorentz transformation is not identical with the Einstein transformation. The Einsteinian addition of velocities is quite different as also the expression for the "relative" density of electricity.

It is true that the Maxwell-Lorentz field equations remain *practically* unchanged by the Lorentz transformation, but they *are* changed to some slight extent. One marked advantage of the Einstein transformation consists in the fact that the field equations of a moving system preserve *exactly* the same form as those of a stationary system.

It should also be noted that the Fresnelian convection coefficient comes out in the theory of relativity as a direct consequence of Einstein's addition of velocities and is quite independent of any electrical theory of matter.

[P. C. M.]

Note 3.

See Lorentz, Theory of Electrons (English edition), § 181, page 213.

H. Poincare, Sur la dynamique 'electron, Rendiconti del circolo matematico di Palermo 21 (1906).

[P. C. M.]

Note 4.
Relativity Theorem and Relativity-Principle.

Lorentz showed that the Maxwell-Lorentz system of electromagnetic field-equations remained practically unchanged by the Lorentz transformation. Thus the electromagnetic laws of Maxwell and Lorentz *can be definitely proved* "to be independent of the manner in which they are referred to two coordinate systems which have a uniform translatory motion relative to each other." (See "Electrodynamics of Moving Bodies," page 5.) Thus so far as the electromagnetic laws are concerned, the principle of relativity *can be proved to be true.*

But it is not known whether this principle will remain true in the case of other physical laws. We can always proceed on the assumption that it does remain true. Thus it is always possible to construct physical laws in such a way that they retain their form when referred to moving coordinates. The ultimate ground for formulating physical laws in this way is merely a subjective conviction that the principle of relativity is universally true. There is no *a priori* logical necessity that it should be so. Hence the Principle of Relativity (so far as it is applied to phenomena other than electromagnetic) must be regarded as a *postulate*, which we have assumed to be true, but for which we cannot adduce any definite proof, until after the generalisation is made and its consequences tested in the light of actual experience.

[P. C. M.]

Note 5.

See "Electrodynamics of Moving Bodies," p. 5-8.

Note 6.
Field Equations in Minkowski's Form.

Equations (i) and (ii) become when expanded into Cartesians:—

$$\partial m_z/\partial y - \partial m_y/\partial z - \partial e_x/\partial \tau = \varrho v_x \}$$

$$\partial m_x/\partial z - \partial m_z/\partial x - \partial e_y/\partial \tau = \varrho v_y \} \dots (1\cdot 1)$$

$$\partial m_y/\partial x - \partial m_x/\partial y - \partial e_z/\partial \tau = \varrho v_z \}$$

and $\partial e_x/\partial x + \partial e_y/\partial y + \partial e_z/\partial z = \varrho$ (2·1)

Substituting x_1, x_2, x_3, x_4 and x, y, z, and $i\tau$; and $\varrho_1, \varrho_2, \varrho_3, \varrho_4$ for ϱv_x, ϱv_y, ϱv_z, $i\varrho$, where $i = \sqrt{(-1)}$.

We get,

$$\partial m_z/\partial x_2 - \partial m_y/\partial x_3 - i(\partial e_x/\partial x_4) = \varrho v_x\{ = \varrho_1 \}$$

$$- \partial m_z/\partial x_1 + \partial m_x/\partial x_3 - i(\partial e_y/\partial x_4) = \varrho v_y = \varrho_2 \} \dots (1\cdot 2)$$

$$\partial m_y/\partial x_1 - \partial m_x/\partial x_2 - i(\partial e_z/\partial x_4) = \varrho v_z\{ = \varrho_3 \}$$

and multiplying (2·1) by i we get

$$\partial ie_x/\partial x_1 + \partial ie_y/\partial x_2 + \partial ie_z/\partial x_3 = i\varrho = \varrho_4 \dots \dots (2\cdot 2)$$

Now substitute

$m_x = f_{23} = -f_{32}$ and $ie_x = f_{41} = -f_{14}$

$m_y = f_{31} = -f_{13}$ $ie_y = f_{42} = -f_{24}$

$m_z = f_{12} = -f_{21}$ $ie_z = f_{43} = -f_{34}$

and we get finally:—

$$\partial f_{12}/\partial x_2 + \partial f_{13}/\partial x_3 + \partial f_{14}/\partial x_4 = \varrho_1 \}$$

$$\partial f_{21}/\partial x_1 + \partial f_{23}/\partial x_3 + \partial f_{24}/\partial x_4 = \varrho_2 \} \dots (3)$$

$$\partial f_{31}/\partial x_1 + \partial f_{32}/\partial x_2 + \partial f_{34}/\partial x_4 = \varrho_3 \}$$

$$\partial f_{41}/\partial x_1 + \partial f_{42}/\partial x_2 + \partial f_{43}/\partial x_3 = \varrho_4 \}$$

Note 9.

On the Constancy of the Velocity of Light.

Page 12—refer also to page 6, of Einstein's paper.

One of the two fundamental Postulates of the Principle of Relativity is that the velocity of light should remain constant whether the source is moving or stationary. It follows that even if a radiant source S move with a velocity u, it should always remain the centre of spherical waves expanding outwards with velocity c.

At first sight, it may not appear clear why the velocity should remain constant. Indeed according to the theory of Ritz, the velocity should become $c + u$, when the source of light moves towards the observer with the velocity u.

Prof. de Sitter has given an astronomical argument for deciding between these two divergent views. Let us suppose there is a double star of which one is revolving about the common centre of gravity in a circular orbit. Let the observer be in the plane of the orbit, at a great distance Δ.

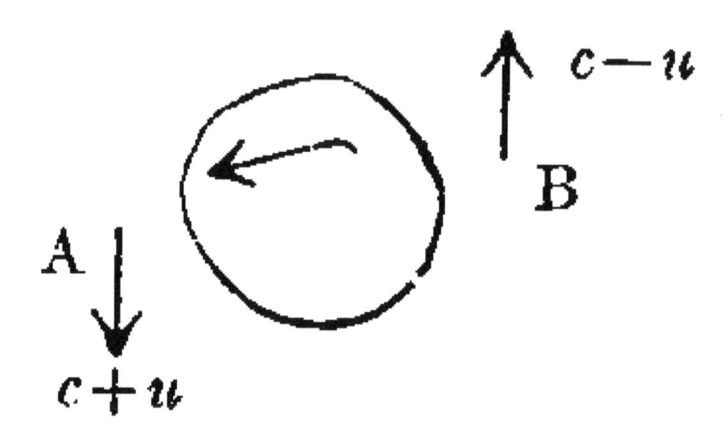

The light emitted by the star when at the position A will be received by the observer after a time, $\Delta/(c + u)$ while the light emitted by the star when at the position B will be received after a time $\Delta/(c - u)$. Let T be the real half-period of the star. Then the observed half-period from B to A is approximately $T - 2\Delta u/c^2$ and from A to B is $T + 2\Delta u/c^2$. Now if $2u\Delta/c^2$ be comparable to T, then it is impossible that the observations should satisfy Kepler's Law. In most of the spectroscopic binary stars, $2u\Delta/c^2$ are not only of the same order as T, but are mostly much larger. For example, if $u = 100$ *km*/sec, T = 8 days, $\Delta/c = 33$ years (corresponding to an annual parallax of ·1"), then $T - 2u\Delta/c^2 = 0$. The existence of the Spectroscopic binaries, and the fact that they follow Kepler's Law is therefore a proof that c is not affected by the motion of the source.

In a later memoir, replying to the criticisms of Freundlich and Günthick that an apparent eccentricity occurs in the motion proportional to $ku\Delta_0$, u_0 being the maximum value of u, the velocity of light emitted being

$u_0 = c + ku$,

$k = 0$ Lorentz-Einstein

$k = 1$ Ritz.

Prof. de Sitter admits the validity of the criticisms. But he remarks that an upper value of k may be calculated from the observations of the double star β-Aurigae. For this star, the parallax $\pi = ·014"$, $e = ·005$, $u_0 = 110$ *km*/sec, T = 3·96,

$\Delta > 65$ light-years,

k is $< ·002$.

For an experimental proof, see a paper by C. Majorana. Phil. Mag., Vol. 35, p. 163.

[M. N. S.]

Note 10.
Rest-density of Electricity.

If ϱ is the volume density in a moving system then $\varrho\sqrt{(1 - u^2)}$ is the corresponding quantity in the corresponding volume in the fixed system, that is, in the system at rest, and hence it is termed the rest-density of electricity.

[P. C. M.]

Note 11

(page 17)

Space-time vectors of the first and the second kind.

As we had already occasion to mention, Sommerfeld has, in two papers on four dimensional geometry (*vide*, Annalen der Physik, Bd. 32, p. 749; and Bd. 33, p. 649), translated the ideas of Minkowski into the language of four dimensional geometry. Instead of Minkowski's space-time vector of the first kind, he uses the more expressive term 'four-vector,' thereby making it quite clear that it represents a directed quantity like a straight line, a force or a momentum, and has got 4 components, three in the direction of space-axes, and one in the direction of the time-axis.

The representation of the plane (defined by two straight lines) is much more difficult. In three dimensions, the plane can be represented by the vector perpendicular to itself. But that artifice is not available in four dimensions. For the perpendicular to a plane, we now have not a single line, but an infinite number of lines constituting a plane. This difficulty has been overcome by Minkowski in a very elegant manner which will become clear later on. Meanwhile we offer the following extract from the above mentioned work of Sommerfeld.

(Pp. 755, Bd. 32, Ann. d. Physik.)

"In order to have a better knowledge about the nature of the six-vector (which is the same thing as Minkowski's space-time vector of the *2nd* kind) let us take the special case of a piece of plane, having unit area (contents), and the form of a parallelogram, bounded by the four-vectors u, v, passing through the origin. Then the projection of this piece of plane on the xy plane is given by the projections u_x, u_y, v_x, v_y of the four vectors in the combination

$$\varphi_{xy} = u_x v_y - u_y v\{x\}.$$

Let us form in a similar manner all the six components of this plane φ. Then six components are not all independent but are connected by the following relation

$$\varphi_{yz}\varphi_{xl} + \varphi_{zx}\varphi_{yl} + \varphi_{xy}\varphi_{zl} = 0$$

Further the contents $|\varphi|$ of the piece of a plane is to be defined as the square root of the sum of the squares of these six quantities. In fact,

$$|\varphi|^2 = \varphi_{yz}^2 + \varphi_{zx}^2 + \varphi_{xy}^2 + \varphi_{xl}^2 + \varphi_{yl}^2 + \varphi_{zl}^2.$$

Let us now on the other hand take the case of the unit plane φ^* normal to φ; we can call this plane the Complement of φ. Then we have the following relations between the components of the two plane:—

$$\varphi_{yz}^* = \varphi_{xl}, \varphi_{zx}^* = \varphi_{yl}, \varphi_{xy}^* = \varphi_{zl} \varphi_{zl}^* = \varphi_{yx} \ldots$$

The proof of these assertions is as follows. Let u^*, v^* be the four vectors defining φ^*. Then we have the following relations:—

$$u_x^* u_x + u_y^* u_y + u_z^* u_z + u_l^* u_l = 0$$

$$u_x^* v_x + u_y^* v_y + u_z^* v_z + u_l^* v_l = 0$$

$$v_x^* u_x + v_y^* u_y + v_z^* u_z + v_l^* u_l = 0$$

$$v_x^* v_x + v_y^* v_y + v_z^* v_z + v_l^* v_l = 0$$

If we multiply these equations by v_l, u_l, v_s, and subtract the second from the first, the fourth from the third we obtain

$$u_x^* \varphi_{xl} + u_y^* \varphi_{yl} + u_z^* \varphi_{zl} = 0$$

$$v_x^* \varphi_{zl} + v_y^* \varphi_{yl} + v_z^* \varphi_{zl} = 0$$

multiplying these equations by $v_x^* . u_x^*$, or by $v_y^* . u_y^*$, we obtain

$$\varphi_{xz}^* \varphi_{xl} + \varphi_{yz}^* \varphi_{yl} = 0 \text{ and } \varphi_{xy}^* \varphi_{xl} + \varphi_{zx}^* \varphi_{zl} = 0$$

from which we have

$$\varphi_{yz}^* : \varphi_{xy}^* : \varphi_{zx}^* = \varphi_{xl} : \varphi_{zl} : \varphi_{yl}$$

In a corresponding way we have

$$\varphi_{yz} : \varphi_{xy} : \varphi_{zx} = \varphi_{xl}^* : \varphi_{zl}^* : \varphi_{yl}^*.$$

i.e. $\varphi_{ik}^* = \lambda\varphi_{(ik)}$

when the subscript (ik) denotes the component of φ in the plane contained by the lines other than (ik). Therefore the theorem is proved.

We have $(\varphi\, \varphi^*) = \varphi_{yz} \varphi_{yz}^* + \ldots$

$= 2\, (\varphi_{yz} \varphi_{zl} + \ldots)$

$= 0$

The general six-vector f is composed from the vectors φ, φ^* in the following way:—

$$f = \varrho\varphi + \varrho^* \varphi^*,$$

ϱ and ϱ^* denoting the contents of the pieces of mutually perpendicular planes composing f. The "conjugate Vector" f^* (or it may be called the complement of f) is obtained by interchanging ϱ and ϱ^*.

We have

$$f^* = \varrho^*\varphi + \varrho\varphi^*$$

We can verify that

$$f_{yz}^* = f_{xl} \text{ etc.}$$

and $f^2 = \varrho^2 + \varrho^{*2}$, $(ff^*) = 2\varrho\varrho^*$.

$| f |^2$ and (ff^*) may be said to be invariants of the six vectors, for their values are independent of the choice of the system of co-ordinates.

[M. N. S.]

Note 12.
Light-velocity as a maximum.

Page 23, and Electro-dynamics of Moving Bodies, p. 17.

Putting $v = c - x$, and $w = c - \lambda$, we get

$$V = (2c - (x + \lambda))/(1 + (c - x)(c - \lambda)/c^2) = (2c - (x + \lambda))/(c^2 + c^2 - (x + \lambda)c + x\lambda/c^2)$$

$$= c\,(2c - (x + \lambda))/(2c - (x + \lambda) + x\lambda/c)$$

Thus v lt; c, so long as $| x\lambda | > 0$.

Thus the velocity of light is the absolute maximum velocity. We shall now see the consequences of admitting a velocity $W > c$.

Let A and B be separated by distance l, and let velocity of a "signal" in the system S be $W > c$. Let the (observing) system S' have velocity $+v$ with respect to the system S.

Then velocity of signal with respect to system S' is given by $W' = (W - v)/(1 - Wv/c^2)$

Thus "time" from A to B as measured in S', is given by $l/W' = l(1 - Wv/c^2)/(W - v) = t'$ (1)

Now if v is less than c, then W being greater than c (by hypothesis) W is greater than v, *i.e.*, $W > v$.

Let $W = c + \mu$ and $v = c - \lambda$.

Then $Wv = (c + \mu)(c - \lambda) = c^2 + (\mu + \lambda)c - \mu\lambda$.

Now we can always choose v in such a way that Wv is greater than c^2, since Wv is $> c^2$ if $(\mu + \lambda)c - \mu\lambda$ is > 0, that is, if $\mu + \lambda > \mu\lambda/c$; which can always be satisfied by a suitable choice of λ.

Thus for $W > c$ we can always choose λ in such a way as to make $Wv > c^2$, *i.e.*, $\lambda - Wv/c^2$ negative. But $W - v$ is always positive. Hence with $W > c$, we can always make t', the time from A to B in equation (1) "negative." That is, the signal starting from A will reach B (as observed in system S') in less than no time. Thus the effect will be perceived before the cause commences to act, *i.e.*, the future will precede the past. Which is absurd. Hence we conclude that $W > c$ is an impossibility, there can be no velocity greater than that of light.

It is *conceptually* possible to imagine velocities greater than that of light, but such velocities cannot occur in reality. Velocities greater than c, will not produce any effect. Causal effect of any physical type can never travel with a velocity greater than that of light.

[P. C. M.]

Notes 13 and 14.

We have denoted the four-vector ω by the matrix $| \omega_1 \; \omega_2 \; \omega_3 \; \omega_4 |$. It is then at once seen that $[=\omega]$ denotes the reciprocal matrix

$$\begin{vmatrix} \omega_1 \\ \omega_2 \\ \omega_3 \\ \omega_4 \end{vmatrix}$$

It is now evident that while $\omega^1 = \omega A$, $[=\omega]^1 = A^{-1}[=\omega]$

$[\omega, s]$ The vector-product of the four-vector ω and s may be represented by the combination

$$[\omega s] = [=\omega]s - \square\omega$$

It is now easy to verify the formula $f^1 = A^{-1}fA$. Supposing for the sake of simplicity that f represents the vector-product of two four-vectors ω, s, we have

$$f' = [\omega^1 s^1] = [[=\omega]^1 s^1 - [=s]^1\omega^1]$$

$$= [A^{-1} [=\omega]sA - A^{-1}s[=\omega]A]$$

$$= A^{-1}[[=\omega]s - s[=\omega]]A = A^{-1}fA.$$

Now remembering that generally

$$f = \varrho\varphi + \varrho^*\varphi^*.$$

Where ϱ, ϱ^* are scalar quantities, φ, φ^* are two mutually perpendicular unit planes, there is no difficulty in seeming that

$$f^1 = A^{-1}fA.$$

Note 15.
The vector product (wf). (P. 36).

This represents the vector product of a four-vector and a six-vector. Now as combinations of this type are of frequent occurrence in this paper, it will be better to form an idea of their geometrical meaning. The following is taken from the above mentioned paper of Sommerfeld.

"We can also form a vectorial combination of a four-vector and a six-vector, giving us a vector of the third type. If the six-vector be of a special type, *i.e.*, a piece of plane, then this vector of the third type denotes the parallelopiped formed of this four-vector and the complement of this piece of plane. In the general case, the product will be the geometric sum of two parallelopipeds, but it can always be represented by a four-vector of the 1st type. For two pieces of 3-space volumes can always be added together by the vectorial addition of their components. So by the addition of two 3-space volumes, we do not obtain a vector of a more general type, but one which can always be represented by a four-vector (loc. cit. p. 759). The state of affairs here is the same as in the ordinary vector calculus, where by the vector-multiplication of a vector of the first, and a vector of the second type (*i.e.*, a polar vector), we obtain a vector of the first type (axial vector). The formal scheme of this multiplication is taken from the three-dimensional case.

Let $A = (A_x, A_y, A_z)$ denote a vector of the first type, $B = (B_{yz}, B_{zx}, B_{xy})$ denote a vector of the second type. From this last, let us form three special vectors of the first kind, namely—

$$\left.\begin{array}{l} B_x = (B_{xx}, B_{xy}, B_{xz}) \\ B_y = (B_{yx}, B_{yy}, B_{yz}) \\ B_z = (B_{zx}, B_{zy}, B_{zz}) \end{array}\right\} \quad (B_{ik} = - B_{ki}, B_{ii} = 0).$$

Since B_{jj} is zero, B_j is perpendicular to the j-axis. The j-component of the vector-product of A and B is equivalent to the scalar product of A and B_j, *i.e.*,

$$(A\, B_j) = A_x B_{jx} + A_y B_{jy} + A_z B_{jz}.$$

We see easily that this coincides with the usual rule for the vector-product; *e. g.*, for $j = x$.

$$(AB_x) = A_y B_{xy} - A_z B_{zx}.$$

Correspondingly let us define in the four-dimensional case the product (Pf) of any four-vector P and the six-vector f. The j-component ($j = x, y, z$, or l) is given by

$$(Pf_j) = P_x f_{jx} + P_y f_{jy} + P_w f_{jz} + P_z f_{jl}$$

Each one of these components is obtained as the scalar product of P, and the vector f_j which is perpendicular to j-axis, and is obtained from f by the rule $f_j = [(f_{jx}, f_{jy}, f_{jz}, f_{jl}) f_{jj} = 0.]$

We can also find out here the geometrical significance of vectors of the third type, when $f = \varphi$, *i.e.*, f represents only one plane.

We replace φ by the parallelogram defined by the two four-vectors U, V, and let us pass over to the conjugate plane φ^*, which is formed by the perpendicular four-vectors U^*, V^*. The components of $(P\varphi)$ are then equal to the 4 three-rowed under-determinants $D_x\ D_y\ D_z\ D_l$ of the matrix

$$\begin{vmatrix} P_x & P_y & P_z & P_l \\ U_x^* & U_y^* & U_z^* & U_l^* \\ V_x^* & V_y^* & V_z^* & V_l^* \end{vmatrix}$$

Leaving aside the first column we obtain

$$D_x = P_y(U_z^* V_l^* - U_l^* V_z^*) + P_z(U_l^* V_y^* - U_y^* V_l^*)$$

$$+ P_l(U_y^* V_z^* - U_z^* V_y^*)$$

$$= P_y \varphi_{zy}^* + P_z^* \varphi_{ly} + P_l \varphi^*_{yz}.$$

$$= P_y \varphi_{xy} + P_z \varphi_{xz} + P_l \varphi_{xl},$$

which coincides with $(P\varphi_x)$ according to our definition.

Examples of this type of vectors will be found on page 36, $\Phi = wF$, the electrical-rest-force, and $\psi = 2wf^*$, the magnetic-rest-force. The rest-ray $\Omega = iw[\Phi\psi]^*$ also belong to the same type (page 39). It is easy to show that

$$\Omega = -i \begin{vmatrix} w_1 & w_2 & w_3 & w_4 \\ \Phi_1 & \Phi_2 & \Phi_3 & \Phi_4 \\ \psi_1 & \psi_2 & \psi_3 & \psi_4 \end{vmatrix}$$

When $(\Omega_1, \Omega_2, \Omega_3) = 0$, $w_4 = i$, Ω reduces to the three-dimensional vector

| $\Omega_1, \Omega_2, \Omega_3$ | = | Φ_1 Φ_2 Φ_3 |

| |

| ψ_1 ψ_2 ψ_3 |

Since in this case, $\Phi_1 = w_4 F_{14} = e_n$ (the electric force)

$\psi_1 = -iw_4 f_{23} = m_x$ (the magnetic force)

we have (Ω) = | e_x e_y e_z |

| m_x m_y m_z |

[M. N. S.]

Note 16.

The electric-rest force. (Page 37.)

The four-vector $\varphi = wF$ which is called by Minkowski the electric-rest-force (elektrische Ruh-Kraft) is very closely connected to Lorentz's Ponderomotive force, or the force acting on a moving charge. If ϱ is the density of charge, we have, when $\varepsilon = 1$, $\mu = 1$, *i.e.*, for free space

$$\varrho_0\varphi_1 = \varrho_0[w_1 F_{11} w_2 F_{12} + w_3 F_{13} + w_4 F_{14}]$$

$$= \varrho_0/(\sqrt{(1 - V^2/c^2)})\ [d_x + 1/c\ (v_2 h_3 - v_3 h_2)]$$

Now since $\varrho_0 = \varrho\sqrt{(1 - V^2/c^2)}$

We have $\varrho_0\varphi_1 = \varrho[d_x + 1/c\ (v_2 h_3 - v_3 h_2)]$

N. B.—We have put the components of e equivalent to (d_x, d_y, d_z), and the components of m equivalent to h_x h_y h_z), in accordance with the notation used in Lorentz's Theory of Electrons.

We have therefore

$$\varrho_0\ (\varphi_1, \varphi_2, \varphi_3) = \varrho\ (d + 1/c\ [v \cdot h]),$$

i.e., ϱ_0 (φ_1, φ_2, φ_3) represents the force acting on the electron. Compare Lorentz, Theory of Electrons, page 14.

The fourth component φ_4 when multiplied by ϱ_0 represents i-times the rate at which work is done by the moving electron, for $\varrho_0\ \varphi_4 = i\varrho\ [v_x d_x + v_y d_y + v_z d_z] = v_x\ \varrho_0\varphi_1 + v_y\ \varrho_0\varphi_2 + v_z\ \varrho_0\varphi_3$. $-\sqrt{(-1)}$ times the power possessed by the electron therefore represents the fourth component, or the time component of the force-four-vector. This component was first introduced by Poincare in 1906.

The four-vector $\psi = i\omega F^*$ has a similar relation to the force acting on a moving magnetic pole.

[M. N. S.]

Note 17.

Operator "Lor" (§ 12, p. 41).

The operation | $\partial/\partial x_1$ $\partial/\partial x_2$ $\partial/\partial x_3$ $\partial/\partial x_4$ | which plays in four-dimensional mechanics a rôle similar to that of the operator ($i\partial/\partial x$, + $j\partial/\partial y$, + $k\partial/\partial z = \nabla$) in three-dimensional geometry has been called by Minkowski 'Lorentz-Operation' or shortly 'lor' in honour of H. A. Lorentz, the discoverer of the theorem of relativity. Later writers have sometimes used the symbol □ to denote this operation. In the above-mentioned paper (Annalen der Physik, p. 649, Bd. 38) Sommerfeld has introduced the terms, Div (divergence), Rot (Rotation), Grad (gradient) as four-dimensional extensions of the corresponding three-dimensional operations in place of the general symbol lor. The physical significance of these operations will become clear when along with Minkowski's method of treatment we also study the geometrical method of Sommerfeld. Minkowski begins here with the case of lor S, where S is a six-vector (space-time vector of the 2nd kind).

This being a complicated case, we take the simpler case of lor s,

where s is a four-vector = | s_1, s_2, s_3, s_4 |

and s = | s_1 |

| s_2 |

| s_3 |

| s_4 |

The following geometrical method is taken from Sommerfeld.

Scalar Divergence—Let $\Delta\Sigma$ denote a small four-dimensional volume of any shape in the neighbourhood of the space-time point Q, dS denote the three-dimensional bounding surface of $\Delta\Sigma$, n be the outer normal to dS. Let S be any four-vector, P_n its normal component. Then

Div S = Lim $\int P_n dS/\Delta\Sigma$.

$\Delta\Sigma = 0$

Now if for $\Delta\Sigma$ we choose the four-dimensional parallelopiped with sides (dx_1, dx_2, dx_3, dx_4), we have then

Div S = $\partial s_1/\partial x_1 + \partial s_2/\partial x_2 + \partial s_3/\partial x_3 + \partial s_4/\partial x_4$ = lor S.

If f denotes a space-time vector of the second kind, lor f is equivalent to a space-time vector of the first kind. The geometrical significance can be thus brought out. We have seen that the operator 'lor' behaves in every respect like a four-vector. The vector-product of a four-vector and a six-vector is again a four-vector. Therefore it is easy to see that lor S will be a four-vector. Let us find the component of this four-vector in any direction s. Let S denote the three-space which passes through the point Q (x_1, x_2, x_3, x_4) and is perpendicular to s, ΔS a very small part of it in the region of Q, $d\sigma$ is an element of its two-dimensional surface. Let the perpendicular to this surface lying in the space be denoted by n, and let f_{sn} denote the component of f in the plane of (sn) which is evidently conjugate to the plane $d\sigma$. Then the s-component of the vector divergence of f because the operator lor multiplies f vectorially.

$$= \text{Div } f_s = \text{Lim } (\textstyle\int f_{sn} d\sigma)/\Delta S.$$

$$\Delta s = 0$$

Where the integration in $d\sigma$ is to be extended over the whole surface.

If now s is selected as the x-direction, Δs is then a three-dimensional parallelopiped with the sides dy, dz, dl, then we have

$$\text{Div } f_x = \frac{1}{dy\, dz\, dl} \left\{ dz.\, dl. \frac{\partial f_{xy}}{\partial y} dy + dl\, dy \cdot \frac{\partial f_{xz}}{\partial z} dz \right.$$

$$\left. + dy\, dz \frac{\partial f_{xl}}{\partial l} dl \right\} = \frac{\partial f_{xy}}{\partial y} + \frac{\partial f_{xz}}{\partial z} + \frac{\partial f_{xl}}{\partial l},$$

and generally

$\text{Div } f_j = \partial f_{jx}/\partial x + \partial f_{jy}/\partial y + \partial f_{jz}/\partial z + \partial f_{jl}/\partial l$ (where $f_{j,j} = 0$).

Hence the four-components of the four-vector lor S or Div. f is a four-vector with the components given on page 42.

According to the formulae of space geometry, D_x denotes a parallelopiped laid in the (y-z-l) space, formed out of the vectors (P_y P_z P_l), (U_y^* U_z^* U_l^*) (V_y^* V_z^* V_l^*).

D_x is therefore the projection on the *y-z-l* space of the parallelopiped formed out of these three four-vectors (P, U^*, V^*), and could as well be denoted by Dyzl. We see directly that the four-vector of the kind represented by (D_x, D_y, D_z, D_l) is perpendicular to the parallelopiped formed by (P U^* V^*).

Generally we have

$$(Pf) = PD + P^*D^*.$$

∴ The vector of the third type represented by (Pf) is given by the geometrical sum of the two four-vectors of the first type PD and P^*D^*.

[M. N. S.]

Footnotes

1. See Note 1.

2. See Note 2.

3. See Note 4.

4. See Notes 9 and 12.

5. Note A.

6. *Vide* Note 9.

7. *Vide* Note 9.

8. *Vide* Note 12.

9. *Vide* Note 1.

10. Note 2.

11. *Vide* Note 3.

12. *Vide* Note 4.

13. Note 5.

14. See notes on § 8 and 10.

15. See note 9.

16. See Note.

17. Vide Note.

18. Just as beings which are confined within a narrow region surrounding a point on a spherical surface, may fall into the error

that a sphere is a geometric figure in which one diameter is particularly distinguished from the rest.

19. Einzelne stelle der Materie.

20. Vide Note.

21. *Vide* note 13.

22. *Vide* note 14.

23. *Vide* note 15.

24. *Vide* note 16.

25. *Vide* note 17.

26. *Vide* note 19.

27. *Vide* note 18.

28. Vide note 40.

29. Sichel—a German word meaning a crescent or a scythe. The original term is retained as there is no suitable English equivalent.

30. Planck, Zur Dynamik bewegter systeme, Ann. d. physik, Bd. 26, 1908, p. 1.

31. H. Minkowski; the passage refers to paper (2) of the present edition.

32. Minkowski—Mechanics, appendix, page 65 of paper (2). Planck—Verh. d. D. P. G. Vol. 4, 1906, p. 136.

33. Schütz, Gött. Nachr. 1897, p. 110.

34. Lienard, L'Eclairage électrique T. 16, 1896, p. 53. Wiechert, Ann. d. Physik, Vol. 4.

35. K. Schwarzschild. Gött-Nachr. 1903. H. A. Lorentz, Enzyklopädie der Math. Wissenschaften V. Art 14, p. 199.

Milton Keynes UK
Ingram Content Group UK Ltd.
UKHW040817051024
449151UK00004B/295